This book examines in detail two of the fundamental questions raised by quantum mechanics. Is the world indeterministic? Are there connections between spatially separated objects?

In the first part of the book after outlining the formalism of quantum mechanics and introducing the measurement problem, the author examines several interpretations, focusing on how each proposes to solve the measurement problem and on how each treats probability. In the second part, the author argues that there can be non-trivial relationships between probability (specifically, determinism and indeterminism) and non-locality in an interpretation of quantum mechanics. The author then reexamines some of the interpretations of part one of the book in the light of this argument, and considers how they fare with regard to locality and Lorentz invariance. One of the important lessons that comes out of this discussion is that any examination of locality, and of the relationship between quantum mechanics and the theory of relativity, should be undertaken in the context of a detailed interpretation of quantum mechanics.

The book will appeal to anyone with an interest in the interpretation of quantum mechanics, including researchers in the philosophy of physics and theoretical physics, as well as graduate students in those fields.

Quantum chance and non-locality

Probability and non-locality in the interpretations of quantum mechanics

Quantum chance and non-locality

Probability and non-locality in the interpretations of quantum mechanics

W. Michael Dickson

Indiana University

CAMBRIDGE
UNIVERSITY PRESS

PUBLISHED BY THE PRESS SYNDICATE OF THE UNIVERSITY OF CAMBRIDGE
The Pitt Building, Trumpington Street, Cambridge, United Kingdom

CAMBRIDGE UNIVERSITY PRESS
The Edinburgh Building, Cambridge CB2 2RU, UK
40 West 20th Street, New York NY 10011–4211, USA
477 Williamstown Road, Port Melbourne, VIC 3207, Australia
Ruiz de Alarcón 13, 28014 Madrid, Spain
Dock House, The Waterfront, Cape Town 8001, South Africa

http://www.cambridge.org

First published 1998
First paperback edition 2005

Typeset in 10pt Monotype Times

A catalogue record for this book is available from the British Library

Library of Congress cataloguing in publication data

Dickson, William Michael, 1968–
Quantum chance and non-locality : probability and non-locality in
the interpretations of quantum mechanics / W. Michael Dickson.
p. cm.
Includes bibliographical references and index.
ISBN 0 521 58127 3 hardback
1. Quantum theory. 2. Physics–Philosophy.
3. Determinism (Philosophy). 4. Chance. I. Title.
QC174.12.D53 1998
530.12–dc21 97-8813 CIP

ISBN 0 521 58127 3 hardback
ISBN 0 521 61947 5 paperback

to my parents

Contents

ix

Contents

Preface

There is a kind of science of everyday phenomena at which we are all experts. We can all predict what will happen when gasoline is thrown on the fire, or when a rock is thrown at the window. None of us is surprised when heated water boils, or when cooled water freezes. These everyday scientific facts come easily.

This everyday science is readily extended to the laboratory, where we learn, for example, that sodium burns yellow, or that liquid helium is very cold. With work, we can learn more complicated facts, involving delicate equipment, and complicated procedures. The result is a kind of science of laboratory phenomena, not different in kind from the science of everyday phenomena.

But what about quantum mechanics? It is, purportedly at least, not about phenomena of the sort mentioned thus far. It is, purportedly at least, not about bunsen burners and cathode ray tubes and laboratory procedures, but about much smaller things — protons, electrons, photons, and so on. What is the relation between the science of quantum mechanics and the science of everyday phenomena, or even the science of laboratory phenomena?

It is no part of my aim to answer this question. However, it will be helpful to note some possibilities. One possibility is that, despite appearances, quantum mechanics really is just about bunsen burners and cathode ray tubes and the like. Perhaps Niels Bohr took such an attitude. (I do not pretend to understand what Bohr wrote, but his name is a convenient label.) He apparently supposed that pieces of laboratory equipment — and everyday objects too — are *outside* the explanatory reach of quantum mechanics. On this reading of Bohr, quantum mechanics does not explain the behavior of these objects in terms of 'quantum objects', but instead describes them *directly*. That is, it describes the relations among them and the results of procedures performed with or on them. On this reading of Bohr, quantum mechanics is just a mathematically sophisticated science of laboratory (and everyday) objects.

But what about protons, electrons, and photons? Are pieces of laboratory equipment not made up of them? Does quantum mechanics not describe their behavior too? Bohr must deny such claims. Instead, he must suppose that terms such as 'proton' do not mean what they seem to mean. The positivists of the first half of this century expended much effort trying to make such a view plausible. They argued that such 'theoretical terms' as 'proton', 'electron', and 'photon' are to be understood as referring not to tiny particles, but to clusters of observations. What quantum mechanics *really* asserts when it says 'a photon is located at the place *x*' is just a set of sentences each of which can be verified by direct observation. (Such sentences are called 'observation-sentences'.) An example of such a sentence is: 'if a photographic plate is placed at *x*, then the plate will show a bright spot'.

The positivists' program of reinterpreting the theoretical terms of science has, by most accounts, failed. There does not seem to be any way to make plausible the claim that when quantum mechanics says 'there is a photon at the place *x*', it *really* means to assert some set of observation-sentences. This failure seems to carry Bohr down with it: there does not seem to be any way to make plausible the claim that, despite appearances, quantum mechanics is *really* only about pieces of laboratory equipment and everyday objects. Quantum mechanics is, it seems, not a science of laboratory objects, but a science of very much smaller things.

Van Fraassen takes a less positivistic view.[1] He says that, at least as far as the *meaning* of the theory is concerned, the relation between quantum mechanics and the science of laboratory objects is just what one would think: quantum mechanics is a theory about very small objects (call them 'quantum objects'); laboratory objects are made of quantum objects; and therefore, quantum mechanics is the basis of our science of laboratory objects. For example, quantum mechanics purports to tell us about how protons, electrons, and neutrons behave. Quantum mechanics says that sodium is made of these. Therefore, quantum mechanics purports to tell us how sodium behaves, for example, when it is burned.

For van Fraassen, then, the theoretical terms of quantum mechanics mean what they appear to mean. When quantum mechanics says 'there is a photon at the place *x*', it means what it says. But for van Fraassen, we are not to *believe* everything that quantum mechanics says. 'I wish merely to be agnostic about the existence of the unobservable aspects of the world described by science', he writes.[2] Hence, although quantum mechanics does make claims that go behind the phenomena, we are not to follow it that far. We ought not to believe that quantum mechanics is telling us how things really are behind the phenomena of laboratory and everyday objects. Instead, we ought to believe that quantum mechanics provides a (more or less) good model of how those phenomena come about — quantum mechanics tells a good story about why sodium burns yellow, but it is just a story.

One can of course go further, following the classical realist: quantum mechanics means what it says, and moreover what it says is (more or less) the truth. The classical realist claims, therefore, that quantum mechanics goes behind the phenomena, and indeed tells us just how things really are behind the phenomena. Sodium burns yellow because it really is made of protons and electrons and neutrons, which behave in a certain way.

Although much could be said about the relative merits of these positions, the concern here is not with which of them we should adopt, but with their application to quantum mechanics. For that purpose, we may ignore the differences between van Fraassen's view and the classical realist's view, and begin with what they have in common: an agreement that quantum mechanics describes the world of our experience in terms of a 'subphenomenal' world, the world of quantum objects. To put it differently, quantum mechanics grounds our effective science of laboratory and everyday objects in terms of a (more) fundamental science of quantum objects.

If quantum mechanics were clearly successful at describing the world of our experience in terms of unobservable objects such as protons, then there would be little need for much of contemporary philosophy of physics. However, quantum mechanics is not thus successful. I do not mean that quantum mechanics is not successful at all. As a science of laboratory objects it is magnificent. (Of course, there remain problems internal to the theory. For example, nobody has a completely satisfactory way of describing gravitational forces in quantum mechanics, but in general, the theory works very well as a science of laboratory objects.) If you want to know what will happen when you shine a laser beam at a polarizer, consult quantum mechanics. If we could only believe that Bohr and the positivists were right, then we could leave it at that. Quantum mechanics could be seen as the best science of laboratory devices that we have had to date.

However, granting that the positivistic view of quantum mechanics is implausible, we must face up to the fact that quantum mechanics has a very difficult time grounding our science of laboratory objects in terms of a science of quantum objects. The problem can be put in many forms — and in chapter 1 the problem will be stated precisely — but one is this way: in order for quantum mechanics to derive the behavior of laboratory objects from the behavior of quantum objects, it must already take the behavior of the laboratory objects for granted. For example, quantum mechanics in its usual form must take for granted that large objects are situated in fairly well-defined regions of space. (The cup is on the table; the train is in the station; and so on.) However, if the science of quantum objects is fundamental, and the science of laboratory objects is derived, then presumably we want the properties of laboratory objects (or, at least, our beliefs about them) to be derived from the properties of quantum objects, rather than to be taken as given. As it stands, quantum mechanics can correctly answer the question 'What are laboratory objects like?' only if we *tell* it the answer. Exactly where quantum mechanics goes wrong will be made clear in chapter 1.

The task of 'interpreting' quantum mechanics, then, is to show how quantum mechanics provides a theory of quantum objects that is capable of grounding our science of laboratory and everyday objects, without taking any part of that science for granted.

In general, it is difficult to say whether a proposed interpretation (of which there are many) succeeds. For example, it is not clear just what we should take the phenomena to be. Must an interpretation predict that the Eiffel Tower really does have a fairly definite location, or need it only predict that whenever one looks for the Eiffel Tower, one will find it to be in a fairly definite location? Or is it acceptable to predict merely that people will *believe* that the Eiffel Tower has a definite location? And must people agree about what its location is, or need they merely believe themselves to agree? One's answers to these questions will depend on what one takes the phenomena of everyday and laboratory objects to be. Different interpretations commit to different accounts of what the phenomena are, and readers may find some interpretations to be more plausible than others for this reason.

However, my aim is not to consider all existing interpretations, much less to evaluate them. Instead, my aim is to use a few interpretations as instruments with which to investigate some questions about quantum objects and their relation to laboratory and everyday objects. More specifically, this book is concerned with probability and non-locality at the level of the quantum objects. Do quantum objects behave deterministically in some sense? Indeterministically? Are there ('non-local') connections among widely separated quantum objects? How do these features of quantum objects relate to features of laboratory and everyday objects, or to our beliefs about them? As soon as we recognize that quantum mechanics goes *behind* the phenomena, we may recognize as reasonable the possibility that the quantum-mechanical world is radically different from the phenomenal world, and the relation between them becomes an open question.

Indeed, it is not clear that the question was ever properly closed, though it was, due to the whims of history, foreclosed. After briefly reviewing some of the mathematics of quantum mechanics — quantum probability theory — in a way that is as free of interpretive assumptions as I can make it, and after saying something about what the problem of interpreting quantum mechanics is, I will turn to a time prior to this foreclosure, when the orthodox view (due largely to Bohr) had not yet been forged. For example, Born was eventually the champion of indeterminism, but much earlier, in the same breath that he introduced probabilities to quantum mechanics, he also recognized the serious possibility of a fundamental determinism. This brief lesson from history will open up some possibilities for interpretation.

In the rest of part 1 (chapters 2–5) I consider some of these possibilities as they are found in some existing interpretations. In part 2 (chapters 6–9), I raise questions about locality. First, in chapters 6 and 7, I try to get a handle on just what kinds of 'locality' there are, what kinds are important, and how they are related to determinism and indeterminism. In chapter 8, I consider what conclusions one might draw from the failure of the locality conditions of chapters 6 and 7. In chapter 9, I return to the interpretations of part 1 in the light of the discussion of chapters 6, 7, and 8.

In many ways, the two parts of the book are somewhat independent. However, one of the underlying themes of the book is that questions about determinism and (especially) locality are best addressed in the context of a well-defined interpretation of quantum mechanics. Abstract analysis (such as can be found in chapters 6–8) can go only so far in helping one to understand non-locality, and then the concrete physical details of a given interpretation become important. This point comes to the fore in chapter 9, where we will see that different interpretations answer questions about locality differently.

Although this book does not pretend to be a popular account, I have tried to make it as accessible as possible, given the nature of the topic. For much of the material, readers will need to know very little quantum mechanics or mathematics. Most of the proofs of the theorems that I present in the text have already been published elsewhere in easily available journals, and I have therefore not repeated the proofs here.[3] Short proofs of minor results sometimes appear in the text or in the endnotes. I have also relegated most of the scholarly comments (acknowledgements, hedges, references, and so on) to endnotes, where they are more at home anyway.

Giving thanks, however, is not a scholarly comment; it is good manners, and a pleasure besides. The investigation as given here would have been far less adequate had it not been for the help of many people. I am grateful to them for useful discussions about the foundations of quantum mechanics and thoughtful comments on my work. In particular, I thank David Albert, Jeeva Anandan, Frank Arntzenius, Guido Bacciagaluppi, Jeff Barrett, Joseph Berkowitz, Rob Brosnan, Harvey Brown, Jeffrey Bub, Tim Budden, Jeremy Butterfield, Rob Clifton, Diarmuid Crowley, Eric Curiel, Dennis Dieks, Matthew Donald, Andrew Elby, Michael Friedman, Judy Hammett, Richard Healey, Geoffrey Hellman, Meir Hemmo, R.I.G. Hughes, Jon Jarrett, Martin Jones, J.B. Kennedy, Andrew Lenard, David Malament, James Mattingly, Fred Muller, Phillip Pearle, Itamar Pitowsky, Michael Redhead, Nick Reeder, Simon Saunders, Howard Stein, Charles Twardy, Pieter Vermaas, and Linda Wessels. No doubt there are others I should thank as well, and to them I apologize for my faulty memory. I am also grateful for comments from audiences willing to put up with my half-baked ideas at Bielefeld (Quantum Theory Without Observers), the University of Cambridge, Cleveland (Philosophy of Science Association meeting, 1996), Drexel University (Workshop on the Classical Limit), Indiana University, the University of Minnesota (Workshop on the Quantum Measurement Problem), New Orleans (Philosophy of Science Association meeting, 1994), the University of Notre Dame, the University of Oxford, and the University of Utrecht (Conference on the Modal Interpretation). Material support I was happy and grateful to receive from the University of Notre Dame, the Mellon Foundation for the Humanities, and the International Center for Theoretical Physics. I am especially grateful to Michael Friedman and Indiana University for supporting a year of research that was essential to finishing the book. I owe thanks and much more to Michael Redhead and Jeremy Butterfield for inviting me for an extended visit to the University of Cambridge. The people there have a lot to do with whatever is good about this book. I also owe a special gratitude to James Cushing, who commented extensively on early drafts, and whose role in turning my barely formed thoughts into coherent ideas cannot be overemphasized. Finally, I thank my wife, Misty, who somehow put up with me for measure-one of the time.

Indiana University

M. Dickson

Acknowledgement

Some parts of the this book were adapted from earlier publications, and I am grateful to the publishers for permission to use that material here. Parts of section 2.3 first appeared in *Foundations of Physics* as Dickson (1994b). Parts of section 4.2.5 first appeared in *Philosophy of Science* as Dickson (1996b). Parts of section 5.2.2 first appeared in *Studies in History and Philosophy of Modern Physics* as Dickson (1996c). Various parts of chapters 6 and 7 first appeared in *Synthese* as Dickson (1996a). Parts of section 9.4.1 first appeared in *Bohmian Mechanics and Quantum Theory: An Appraisal* as Dickson (1996d).

Part one

Quantum chance

1

Quantum probability and the problem of interpretation

1.1 Quantum probability and quantum mechanics

1.1.1 The formalism of quantum probability theory

Discussions of quantum mechanics are often confused by a lack of clarity about what exactly constitutes 'quantum mechanics'. It is therefore useful to try at the start to isolate a consistent mathematical core of quantum mechanics, and consider anything that goes beyond this core to be 'interpretation'. For us, this core is quantum probability theory.

Quantum probability is a generalization of classical probability, and therefore I begin with a brief review of the latter. I assume that the reader has some familiarity with the ideas of probability theory. What follows is just to provide a quick review, and to establish some notation and terminology.[1]

In modern classical probability theory, probabilities are defined over algebras of events. The motivation is straightforward: we begin with a set of 'primitive', or 'simple', events (the 'sample space'), and form an algebra of events by taking all logical combinations of the simple events. For example, let us take the simple events to be the possible results of rolling a six-sided die one time, so that the sample space is the set $\{1, 2, 3, 4, 5, 6\}$. We then form an algebra of events from the sample space by taking all possible logical combinations of the simple events. Logical combinations include, for example, 'either 3 or 5' and 'not 3 and not 2'.

In classical probability theory, we represent logical combinations with the set-theoretic operations of intersection (which represents 'and'), union (which represents 'or'), and complement (which represents 'not'). Events are therefore given by sets whose elements are taken from the sample space. The event 'either 3 or 5' is represented by $\{3\} \cup \{5\}$, which is $\{3, 5\}$. The event 'not 3 and not 2' is represented by $\neg\{3\} \cap \neg\{2\}$, which is $\{1, 2, 4, 5, 6\} \cap \{1, 3, 4, 5, 6\}$, which is $\{1, 4, 5, 6\}$. (More precisely, we form an algebra of events from a sample space by closing the sample space under complement and countable union. Doing so guarantees closure under countable intersection.)

We introduce probabilities to the picture with a probability measure, p, over the simple events. This measure is extended to the entire algebra of events by Kolmogorov's axioms:

Kolmogorov's axioms:

(1) $p(\emptyset) = 0$,
(2) $p(\neg F) = 1 - p(F)$,
(3) $p(F \cup F') = p(F) + p(F') - p(F \cap F')$

where F and F' are events in the algebra and $\emptyset$ is the empty set (the set with no elements). In our example, the probability measure over the simple events is given by $p(\{i\}) = 1/6$ for all i. Hence, for example, $p(\{3\} \cup \{5\}) = 1/3$ (by axioms (3) and (1)), and so on.

Finally, we define a conditional probability measure, the probability of some event *given* the occurrence of some other event. For example, we may want to know the probability that the die shows 6 given that it shows either 2 or 6. Conditional probabilities are defined by

$$p(F|F') = \frac{p(F \cap F')}{p(F')}. \tag{1.1}$$

For example, $p(\{6\}|\{2, 6\}) = 1/2$.

To summarize, we may identify a classical probability theory with an ordered triple, $\langle \Omega, \mathscr{F}, p \rangle$, where Ω is the sample space, $\mathscr{F}$ is the algebra of events generated by Ω, and p is a Kolmogorovian probability measure.

Quantum probability theory also begins with an ordered triple, $\langle \mathscr{H}, L_{\mathscr{H}}, \psi \rangle$. Here $\mathscr{H}$ is a Hilbert space, which is a (complete, complex) vector space with an inner product defined on it. (We also require that it have a countable basis.) Every (normalized) vector — or equivalently, every ray — in $\mathscr{H}$ corresponds to a simple event, so that $\mathscr{H}$ may be considered the sample space. We generate an algebra of events, $L_{\mathscr{H}}$, from $\mathscr{H}$ as follows. Beginning with the rays, i.e., the one-dimensional subspaces of $\mathscr{H}$, close under the operations of span, intersection, and orthogonal complement. (The span of two subspaces, P and P', is the set of all vectors that can be written as a weighted sum of vectors from P and P'. For example, the span of two one-dimensional subspaces (rays) is the plane containing both of them. The intersection of two subspaces is the largest set of vectors contained in both of them. The orthogonal complement, or orthocomplement, of a subspace is the largest subspace entirely orthogonal (perpendicular) to it.) These operations correspond to the lattice-theoretic operations of join (denoted '$\vee$'), meet (denoted '∧'), and orthocomplement (denoted '⊥'), respectively, which leads some to interpret them as the quantum-mechanical representation of the logical operations of or, and, and not. (I will discuss this idea further in section 4.1.)

The algebra of quantum-mechanical events is denoted by $L_{\mathcal{H}}$ because it forms a lattice, a partially ordered set for which the operations meet, join, and orthocomplement are defined between each pair of elements. The partial ordering is given by subspace inclusion. Alternatively, the algebra of quantum-mechanical events can be considered to be a partial Boolean algebra. I will discuss this alternative in chapter 4.

Finally, ψ is a vector in $\mathcal{H}$ with norm 1 (i.e., the inner product of ψ with itself is 1). It generates a probability measure, p^{ψ}, over the sample space $\mathcal{H}$ through the familiar rule:

$$p^{\psi}(\varphi) = |(\psi, \varphi)|^2,\tag{1.2}$$

where $(\cdot, \cdot)$ is the inner product. Or, using Dirac notation (which I will use from now on),

$$p^{\psi}(\varphi) = |\langle \psi | \varphi \rangle|^2.$$

Often I will speak of the elements of the sample space not as vectors, but as projections, or subspaces. Every vector can be represented, for present purposes, as the (one-dimensional) subspace that it spans. Also, I will often use the terms 'projection operator' and 'subspace' interchangeably (for there is a one-to-one correspondence between them), and I will use the same notation for both. I will even, at times, say things like 'the projection P is contained in the projection P'', meaning that the subspace onto which P projects is a subspace of the subspace onto which P' projects. None of this loose talk should cause confusion.

Now the story gets a bit more complex. It would be nice if Kolmogorov's axioms held in quantum probability (substituting the lattice-theoretic operations for the set-theoretic ones, of course). Axioms 1 and 2 do hold, but axiom 3 fails in general, though it holds when the events are orthogonal (more precisely, when the subspaces representing the events are orthogonal). That is, we have:

(1) $p(\mathbf{0}) = 0$,
(2) $p(P^{\perp}) = 1 - p(P)$,
(3) $p(P \vee P') = p(P) + p(P')$, when $P \perp P'$,

where $\mathbf{0}$ is the zero subspace (the zero element of the lattice).[2]

These axioms are somewhat unsatisfactory, because they do not, by themselves, tell us how to calculate the probability of events $P \vee P'$ for arbitrary P and P'. Some have argued that we should take this limitation to be a *lesson*: the probability of such events is just undefined. However, the point of my starting with quantum probability is to provide as neutral a basis for interpretation as possible. Hence we may just as well allow into the theory probabilities for such combinations of events, and then later, if we like, remove them.

Moreover, quantum probability as introduced thus far lacks generality in another sense. There exist probability measures on Hilbert spaces that are not representable by a vector through equation (1.2). To capture all of the probability measures over a Hilbert space, we need to represent them not by vectors, but by the so-called density operators,[3] which are (bounded, positive) operators on the Hilbert whose trace is 1. The trace of an operator, W, is given by

$$\mathsf{Tr}[W] = \sum_i \langle \varphi_i | W | \varphi_i \rangle,$$

where $\{|\varphi_i\rangle\}$ is any orthonormal basis for $\mathcal{H}$ — the value of $\mathsf{Tr}[W]$ is independent of the choice of the basis, $\{|\varphi_i\rangle\}$. Hence we alter the definition of a quantum probability theory, so that it consists of an ordered triple, $\langle \mathcal{H}, \mathsf{L}_{\mathcal{H}}, W \rangle$, where W is a density operator, and it generates a probability measure over all of $\mathsf{L}_{\mathcal{H}}$ by

$$p^W(P) = \mathsf{Tr}[WP]. \tag{1.3}$$

Notice that we have simply bypassed the method of classical probability theory: rather than extending a measure over the sample space to a measure over all events through axioms such as Kolmogorov's, we define the measure over $\mathsf{L}_{\mathcal{H}}$ directly, through (1.3). It becomes a matter for investigation what the properties of the measure generated by W are. As it happens, axioms (1)–(3) as stated for our original version of quantum probability theory hold here as well. (The differences are that now: (a) we have a theory that includes all probability measures over the sample space, and (b) we have a theory that tells how to calculate the probability of every event in $\mathsf{L}_{\mathcal{H}}$ directly. We could have gotten (b) without moving to the formalism of density operators, however.)

A *random variable* on a classical probability space is a map from simple events to real numbers. A probability measure over the events therefore induces a probability measure over the *range* of the random variables. In quantum probability theory, random variables are represented by self-adjoint operators on $\mathscr{H}$. Such operators can be conceived as maps from *some* rays in $\mathscr{H}$ to real numbers. Recall that every self-adjoint operator, A, has eigenvectors, $|a_i\rangle$, each of which corresponds to some eigenvalue, a_i. Hence A can be conceived as a map from its eigenvectors to their corresponding eigenvalues.

The set of all eigenvectors of A corresponding to a given eigenvalue, a_i, forms a subspace of $\mathscr{H}$, and may be denoted $P_{a_i}^A$. The set $\{P_{a_i}^A\}$ for all eigenvalues, a_i, of A is a set of mutually orthogonal subspaces spanning $\mathscr{H}$, so that for any density operator, W, $p^W(\bigvee_i P_{a_i}^A) = 1$, and the linearity of the trace functional plus the orthogonality of the $P_{a_i}^A$ further guarantee that the usual sum rule for probabilities holds: $p^W(P_{a_i}^A \vee P_{a_j}^A) = p^W(P_{a_i}^A) + p^W(P_{a_j}^A)$ for $i \neq j$. Hence p^W generates a probability measure over the set of all eigenvalues of A.

Finally, the conditional probability of P given P' in quantum probability is

$$p^W(P|P') = \frac{\mathrm{Tr}[P'WP'P]}{\mathrm{Tr}[WP']}. \tag{1.4}$$

This definition of conditional probabilities is sometimes called 'Lüders' rule'.[4] It is the only definition (given certain constraints) that meets the reasonable criterion that whenever P is contained in P', the conditional probability is given by[5]

$$p^W(P'|P) = \frac{p^W(P')}{p^W(P)}. \tag{1.5}$$

Quantum probability theory is a generalization of classical probability theory.[6] Therefore, not everything that is true in classical probability theories will be true of the more general quantum probability theories. We have already seen one example, in the failure of Kolmogorov's third axiom. Another important difference is that while joint probabilities (probabilities for arbitrary sets of events to be jointly occurrent) are always definable in a classical probability theory, in quantum probability it is not possible to define a joint probability measure for arbitrary sets of events (given some plausible assumptions about joint probabilities).[7] To put it differently: if you pick an arbitrary set of events from $L_{\mathscr{H}}$, you are not guaranteed that there is any probability for this set of events to be jointly occurrent. We may put it yet another way: while joint probability distributions for pairs of random variables always exist in classical probability theory, they need not exist for pairs of operators (more precisely, for the sets of their eigenvalues) in quantum probability theory.

1.1.2 From quantum probability to quantum mechanics

Quantum probability theory is a consistent mathematical theory, but as yet has nothing to do with physics. I have given a few hints about how some relation might be made between quantum probability and physics — in particular, I noted that quantum probability measures can be interpreted as probability measures over all of the eigenvalues of each operator. However, that fact is not enough to generate a physical theory, even after we make the standard identification between operators and physical quantities, i.e., 'observables', so that the eigenvalues of an operator are the possible values of the corresponding observable.

The problem is that it is not at all clear how to get from quantum probability to a consistent and satisfactory physical theory. This problem will occupy us for parts of the next four chapters. To make the problem clear, I begin with a minimal extension of quantum probability, to arrive at the theory that I will call 'quantum mechanics'. Even this minimal extension has its difficulties, as I discuss in the next section. Nonetheless, by making the extension and exposing the problem, we will at least have a handle on the difficulties that face quantum mechanics.

The extension may be given first in the more familiar terms of vectors in Hilbert space. In these terms, the state of a quantum system is represented by a vector, $|\psi\rangle$. A system's state evolves in time according to Schrödinger's equation:

$$\frac{d|\psi(t)\rangle}{dt} = -i\hbar H|\psi(t)\rangle, \tag{1.6}$$

where H is the Hamiltonian operator for a given system. The state of a system at any time generates a probability measure over all possible values of each observable in the way already described.

More generally, the state of a physical system is given by a density operator on Hilbert space. The evolution of a density operator is easily derived from the Schrödinger equation. The result is that the state, $W(t)$, of a system evolves according to a unitary operator, $U(t)$:[8]

$$W(t) = U(t)W(0)U^{-1}(t), \tag{1.7}$$

where $U(t) = e^{-iHt}$. The probability measure over all possible values of each observable is given at each time, t, by $\text{Tr}[W(t)P_a^A]$, where, recall, A represents some observable, and a is some eigenvalue of A. (I shall often not distinguish notationally between observables and operators.)

'Quantum mechanics' therefore makes two claims that go beyond quantum probability. First, it claims that the state of a system is given by a density operator. Second, it claims that the state evolves according to the unitary operator $U(t) = e^{-iHt}$. These two claims may appear innocent enough, but as we shall see in the next section, they lead to a difficult problem. Solving, or avoiding, this problem is one of the central challenges facing interpreters of quantum mechanics.

1.2 Interpreting quantum mechanics

1.2.1 The 'measurement problem'

The problem that quantum mechanics faces — the 'measurement problem' — is that it sometimes assigns the *wrong* state to some systems.[9] (As we shall see, the name 'measurement problem' is misleading, because it suggests that the problem occurs only when one makes a measurement, whereas the problem is, in fact, generic.)

The problem is best described by way of illustration. Suppose that a quantum system begins in the state $|\alpha_1\rangle$, an eigenvector of A with eigenvalue a_1. Suppose we perform a measurement of A, as follows: the measuring device begins in a ready-to-measure state, $|M_0\rangle$, and, after the measurement, is perfectly correlated with the value of A possessed by the system. We may represent the measurement schematically by (assuming for simplicity that the interaction does not disturb the measured system)

initial state   measurement   final state
interaction

$|\alpha_1\rangle|M_0\rangle$ $\longrightarrow$ $|\alpha_1\rangle|M_1\rangle$,

where $|M_1\rangle$ is the state of the apparatus that indicates a value of a_1. (Juxtaposition of two vectors represents a tensor product. Readers less familiar with the tensor product formalism may read juxtaposition as 'and'.[10] For example, read '$|\alpha_1\rangle|M_0\rangle$' as 'the measured system is in the state $|\alpha_1\rangle$ *and* the apparatus is in the state $|M_0\rangle$'.) Similarly, if the quantum system begins in the state $|\alpha_2\rangle$, then the interaction would be

initial state   measurement   final state
interaction

$|\alpha_2\rangle|M_0\rangle$ $\longrightarrow$ $|\alpha_2\rangle|M_2\rangle$,

but now trouble is close at hand. The evolution of the pair of systems during these measurement-interactions must be described by some unitary operator,

$U(t)$, and $U(t)$ is always linear, which means that for any vectors $|\varphi_1\rangle$ and $|\varphi_2\rangle$,

$$U(t)[c_1|\varphi_1\rangle + c_2|\varphi_2\rangle] = c_1 U(t)|\varphi_1\rangle + c_2 U(t)|\varphi_2\rangle, \tag{1.8}$$

where c_1 and c_2 are any complex numbers. Applying (1.8) to the measurement-interactions described above, we get that if the quantum system begins in the state $c_1|\alpha_1\rangle + c_2|\alpha_2\rangle$, then the measurement-interaction yields

initial state	measurement interaction	final state
$(c_1|\alpha_1\rangle + c_2|\alpha_2\rangle)|M_0\rangle$	$\longrightarrow$	$c_1|\alpha_1\rangle|M_1\rangle + c_2|\alpha_2\rangle|M_2\rangle.$

It is not at all clear what to say about the final state in this interaction. What is clear is that when we perform the experiment, we find the apparatus in either the state $|M_1\rangle$ or the state $|M_2\rangle$. Yet, the final state assigned by quantum mechanics is neither of these. Indeed, it is apparently not the sort of state that we ever witness — a 'superposition' of $|\alpha_1\rangle|M_1\rangle$ and $|\alpha_2\rangle|M_2\rangle$. It appears that the standard theory fails: the final state that it assigns to the system (or, the event that is occurrent with probability 1) is one that we never actually see when we perform the experiment. What we see is either $|M_1\rangle$ or $|M_2\rangle$, but quantum mechanics predicts something else entirely.

I have described the 'measurement problem' in the context of a measurement, but the problem is general. It seems likely that the sort of interaction that led quantum mechanics to attribute the 'wrong' state to the measuring apparatus could occur also in situations that we would not call 'measurements'. Indeed, quantum mechanics appears to face the very general problem of not adequately describing the world as we actually see it. The states that it attributes to macroscopic objects are not the states that we observe them to have. Quantum mechanics does a good job of describing the world *behind* the phenomena of our everyday experience, but it appears to fail miserably to describe our everyday experience itself.

1.2.2 Are the quantum probabilities epistemic?

The following line of thought might already have occurred to the reader: Why not suppose that the probabilities that the standard view prescribes are merely epistemic probabilities? That is, why not suppose that when the standard view says that the final state is $c_1|\alpha_1\rangle|M_1\rangle + c_2|\alpha_2\rangle|M_2\rangle$ all it *means* is that one or the other of $|\alpha\rangle|M_1\rangle$ and $|\alpha_2\rangle|M_2\rangle$ is occurrent, with probabilities $|c_1|^2$ and $|c_2|^2$, respectively? (The probabilities $|c_1|^2$ and $|c_2|^2$ are then 'epistemic' because one of the two events is *really* occurrent, but we do not know which — or better, the theory simply does not tell us which.) If we can maintain this interpretation of the quantum probability measure, then the measurement problem apparently disappears.

Of course, the ignorance interpretation faces the problem of explaining what it means for the event given by (the one-dimensional subspace spanned by) $c_1|\alpha_1\rangle|M_1\rangle + c_2|\alpha_2\rangle|M_2\rangle$ to be occurrent with probability 1, but a little metaphysical creativity might produce such an explanation. The real challenge facing this view is that it is not at all clear that the quantum probabilities can be reasonably interpreted as epistemic probabilities. Indeed, the prevailing orthodoxy among physicists (or at least, what is reputed by philosophers of physics to be the prevailing orthodoxy among physicists) is that quantum probabilities are *not* merely epistemic.

Rarely does one find *arguments* for this orthodox view, but arguments do exist. In this section, I consider one argument against the epistemic interpretation of quantum probabilities, and how that argument might be answered.

Consider a quantum system in the state $|\psi\rangle$. (In the formalism of density-operators, the system is in the (pure) state $|\psi\rangle\langle\psi|$.) Consider two events, given by $|\chi\rangle$ and $|\xi\rangle$, where, for $c_1 \neq d_1$ and $|\varphi\rangle\perp|\psi\rangle$,

$$|\chi\rangle = c_1|\psi\rangle + c_2|\varphi\rangle \quad |c_1| > |c_2|,$$
$$|\xi\rangle = d_1|\psi\rangle + d_2|\varphi\rangle \quad |d_1| > |d_2|.$$

(When I say that an event is 'given by' a vector $|\chi\rangle$, I mean that it is represented by the subspace spanned by $|\chi\rangle$.) We therefore have that $p^\psi(|\chi\rangle\langle\chi|) = |c_1|^2$ and $p^\psi(|\xi\rangle\langle\xi|) = |d_1|^2$. Given these probabilities, if you were to adopt an ignorance interpretation of p^ψ then you should be willing to accept the following bets as fair.

Bet 1: If $|\chi\rangle\langle\chi|$ is the truly occurrent event, then you win $|c_2|^2$ dollars, and otherwise you lose $|c_1|^2$ dollars. (The expected value of this bet is $p^\psi(|\chi\rangle\langle\chi|)|c_2|^2 - [1 - p^\psi(|\chi\rangle\langle\chi|)] \times |c_1|^2 = 0$. Hence it is a fair bet.)

Bet 2: If $|\xi\rangle\langle\xi|$ is the truly occurrent event, then you win $|d_2|^2$ dollars, and otherwise you lose $|d_1|^2$ dollars. (The expected value of this bet is $p^\psi(|\xi\rangle\langle\xi|)|d_2|^2 - [1 - p^\psi(|\xi\rangle\langle\xi|)] \times |d_1|^2 = 0$. Hence it is a fair bet.)

For simplicity, assume that if any *other* event occurs, then no money changes hands. Then bets 1 and 2 together form a so-called 'Dutch Book'. That is, although you are committed to agreeing that they are both fair bets, you are guaranteed to lose money if you take both of them (when either $|\chi\rangle\langle\chi|$ or $|\xi\rangle\langle\xi|$ occurs). If $|\chi\rangle\langle\chi|$ occurs, then you get a total of $|c_2|^2 - |d_1|^2$ dollars. If $|\xi\rangle\langle\xi|$ occurs, then you get a total of $|d_2|^2 - |c_1|^2$ dollars. Both of these totals are less than zero. However, the minimum that we should require of an epistemic measure is that it does not commit one to a Dutch Book.

Well, this argument is only as convincing as its hidden assumptions, and there are at least three of those. I now turn to examine these hidden assumptions. Doing so will point to some strategies for avoiding the Dutch Book.

First, the argument assumes that $|\chi\rangle\langle\chi|$ and $|\xi\rangle\langle\xi|$ do not co-occur, but what justification is there for this assumption? After all, we already know that there is no way to assign a joint probability to arbitrary sets of non-orthogonal events, and $|\chi\rangle\langle\chi|$ and $|\xi\rangle\langle\xi|$ are (necessarily) non-orthogonal (because $|c_1| > |c_2|$ and $|d_1| > |d_2|$). Therefore, the assumption that $|\chi\rangle\langle\chi|$ and $|\xi\rangle\langle\xi|$ do not co-occur is not based on their having joint probability zero. One might try to maintain, then, that they *can* co-occur, and doing so ruins the Dutch Book (because then you *can* win both bets).

On the other hand, the intersection of $|\chi\rangle\langle\chi|$ and $|\xi\rangle\langle\xi|$ is the zero subspace — they have nothing in common. Hence, at the very least, if we wish to say that $|\chi\rangle\langle\chi|$ and $|\xi\rangle\langle\xi|$ can co-occur, we cannot adopt the classical ignorance interpretation — we cannot say that one and only one of the events in the sample space is occurrent (which is what we do say in the ignorance interpretation of a classical probability measure). Any proponent of the epistemic interpretation who wishes to avoid the Dutch Book by allowing that $|\chi\rangle\langle\chi|$ and $|\xi\rangle\langle\xi|$ can co-occur must have a story to tell about *how* they can co-occur given that: (1) their lattice-theoretic meet is the zero subspace (i.e., they are *distinct* simple events), and (2) their joint probability is undefined.

The second hidden assumption is that ignorance is represented by the classical 'or'. Allowing that $|\chi\rangle\langle\chi|$ and $|\xi\rangle\langle\xi|$ cannot co-occur, one way to say what led us into a Dutch Book is that we assigned too high a probability to the event '$|\chi\rangle\langle\chi|$ or $|\xi\rangle\langle\xi|$'. We assigned $|c_1|^2$ to $|\chi\rangle\langle\chi|$ and $|d_1|^2$ to $|\xi\rangle\langle\xi|$, and therefore, $|c_1|^2 + |d_1|^2$ to the event '$|\chi\rangle\langle\chi|$ or $|\xi\rangle\langle\xi|$'. However, $|c_1|^2 + |d_1|^2 > 1$, and any time you assign probability greater than 1 to an event, you are immediately committed to a Dutch Book. From this point of view, we can see that if we did not represent our ignorance with the classical 'or' (according to which the probability of 'F or F'' is $p(F) + p(F')$ when F and F' are disjoint events), then we would not (necessarily) be committed to a Dutch Book, because we would not (necessarily) assign a probability greater than 1 to the event '$|\chi\rangle\langle\chi|$ or $|\xi\rangle\langle\xi|$'.

However, merely adopting the lattice-theoretic operations as our logical operations will not, by itself, suffice. We would still apparently be committed to the fairness of bets 1 and 2. Then what reason could we have for refusing to take both bets?

One reason might be a refusal to allow that propositions involving non-orthogonal events are well defined in the first place. This strategy is consonant with the non-existence of a joint probability for arbitrary sets of non-orthogonal events. Following this strategy, we would have justification for agreeing to each of bets 1 and 2, while refusing to agree to both of them together. To agree to both of them together is to take a stand on the joint occurrence or non-occurrence of each of a pair of non-orthogonal events. To put it differently, to agree to both bets together is to take a stand on a statement about non-orthogonal events. Therefore, agreeing to both bets together amounts to betting on the truth or falsity of a statement whose meaning is undefined, and to refuse such a bet seems completely reasonable.

This strategy relies on the impossibility of verifying the occurrence or non-occurrence of each of a pair of non-orthogonal events. Quantum mechanics (at least as it was described in section 1.2) seems to uphold this impossibility. In fact, if we *could* perform an experiment to verify the occurrence or non-occurrence of disjoint and non-orthogonal events, then quantum mechanics would have another serious problem, because it has no way to make predictions about the results of such an experiment.[11]

Nonetheless, proponents of this strategy face at least two difficult challenges. First, they must convince us that ignorance really should be represented by the lattice-theoretic operations rather than the classical ones. Second, they must make plausible the claim that statements involving non-orthogonal events are undefined. (After all, it seems very clear what we mean when we say 'the particle has position x and momentum q', but on the present view, this statement is undefined.) There do, in fact, exist proponents of this view, and later I will discuss whether they have successfully met these and other challenges.

The third hidden assumption behind the Dutch Book argument is that the ignorance measure p^ψ is to be taken as ranging over *all* events in the sample space. Some authors have suggested that we adopt an ignorance interpretation that is restricted to some special set of events. If this set is chosen carefully, then a Dutch Book can be avoided. Of course, proponents of this view must have something to say about the status of events outside the 'special' set. They must also convince us that they solve the measurement problem, by convincing us that the 'special' set of events contains, for example, the final states of the apparatus after the measurement (so that we can properly interpret the measure, p^ψ, over them as an ignorance measure).

I will have more to say about this strategy in chapters 4 and 5, when I discuss modal interpretations and Bohm's theory.

So there are strategies for avoiding the Dutch Book argument, but none of them is as straightforward as it might at first appear. On the other hand, the 'orthodox' view avoids it from the start, by denying the ignorance interpretation of the quantum probability measure. Note that this denial leads to two kinds of indeterminism. First, there is an 'indeterminism of the moment', which might better be called 'indefiniteness': it is not just that the most *we* know about a quantum system is the probability measure, p^ψ; rather, that measure is all that there is to be known. The measure p^ψ *completely* characterizes the quantum system. The endorsement of indefiniteness is, of course, what led my minimal extension of quantum probability into the measurement problem. Second, there is a dynamical indeterminism: because p^ψ completely characterizes the quantum system, there is no way to predict (even in principle) what will be the result of a measurement of an arbitrary observable on the system. To put it differently: the outcome that we witness as the result of the measurement was not determined to occur by anything prior to the measurement.

Let us call these two dogmas 'indefiniteness' and 'dynamical indeterminism'. Each of them has enjoyed sufficient dominance among physicists that we may suppose 'orthodoxy' to require their acceptance. Hence I shall call interpretations that accept both of them 'orthodox interpretations'. However, it would be wrong to suppose that what counts as orthodoxy is anything more than historical accident. To counter any such supposition, some historical therapy is useful. In the next section, I review (briefly) the introduction of probabilities to quantum mechanics. We will see that at the time that Born introduced probabilities to quantum mechanics, it was still recognized that indefiniteness and dynamical indeterminism were not the only options for interpretation.

1.2.3 Fable: A brief history of Born's rule

Prior to the introduction of probabilities, much of quantum physics was concerned with the calculation of so-called stationary states.[12] Two kinds of stationary state were recognized in early quantum theory: the states of constant energy and the states of constant momentum.[13] (An object in a state of stationary energy is one whose energy does not change over time, for example.) It was from this point of view that Born approached the more complex problem of collision, which had been little treated prior to 1926. The problem is this: Given an initial distribution of incident particles (electrons) moving towards a scattering target (an atom), what is the distribution of the scattered particles? Born made the natural assumption that long before the collision, and again long after it, the incident and scattered electrons were in the stationary momentum states corresponding to free electrons, while the scattering target was in a stationary energy state.

Motivated and aided by the recent introduction of Schrödinger's wave mechanics, Born published a preliminary report, in which he claimed that, to first order, the state of the combined system (atom plus electrons) after the scattering was (in modern notation, and ignoring some fine structure):[14]

$$|\Psi_n\rangle = \sum_m \int d\alpha \; c_{nm}(\alpha) \, |\chi_{nm}(\alpha)\rangle |\psi_m\rangle, \qquad (1.9)$$

where $|\chi_{nm}(\alpha)\rangle$ is a stationary momentum state of an electron moving in the direction α (an angle in three-space) that has suffered a change in energy $E_n - E_m$, and $|\psi_m\rangle$ is the stationary energy state of the atom, with energy E_m. (Suppose that it began with energy E_n and that the spectrum of stationary states is characterized by the discrete energy-spectrum $\{E_0, E_1, \ldots\}$.)

How is this result to be interpreted? Mathematically, it is just a 'sum' (superposition) of stationary momentum states for the freely moving electrons correlated with the stationary energy states of the atom. Born probably saw it primarily as a wave — perhaps the natural interpretation of such a superposition. Indeed, contrary to common lore, Born was apparently initially enthusiastic about a Schrödinger-inspired interpretation in terms of waves.[15] However, whatever his initial commitments, Born did as well make this fateful observation:

If one wants to interpret this result [eq. (1.9)] in terms of particles rather than waves, then there is only one interpretation possible: [$c_{nm}(\alpha)$] represents the probability that the electron coming in from the z direction will be thrown into the direction determined by [α] ..., where its energy has increased by a quantum [$E_n - E_m$] at the expense of the atomic energy.[16]

In a footnote added in proof, Born corrected himself, noting that the probability is not given by $c_{nm}(\alpha)$, but instead $|c_{nm}(\alpha)|^2$. Of course, in a suitably generalized form, Born's suggestion is just equation (1.3).

This preliminary report was soon followed by a longer paper in which Born gave some calculations to justify his claim, and further developed the interpretation of (1.9). He had apparently realized by this time that any wavefunction, ψ, can be expressed as a sum of stationary energy states (if the domain of ψ is discrete) or stationary momentum states (if the domain of ψ is continuous). In one dimension:

$$\psi(q) = \sum_n c_n \eta_n(q),$$

(1.10)

$$\psi(x) = 1/2\pi \int dk \, c(k) e^{ikx}$$

(1.11)

(where $k = p/\hbar$). Born wrote:[17] '$|c_n|^2$ denotes [in eq. (1.10)] the frequency of the state n, and the total number [of atoms in a system described by $\psi(q)$] is composed additively of these components'; and later: '$|c(k)|^2$ is [in eq. (1.11)] the frequency of a motion with the momentum $p = (h/2\pi)k$'. Born did not (in this paper) draw any conclusions about 'irreducible indeterminism', nor did he describe anything that one would today recognize as 'Born's interpretation', nor even did he generalize his result to the so-called 'Born rule'. Indeed, Pauli was the one to make the generalization that would bear Born's name.[18] Pauli made the remark in an almost off-hand way, in a footnote. It is worth keeping in mind that equation (1.3), which today we consider to be at the center of the theory, was not greeted with the fanfare that we might expect.

Of course, Born did come to espouse some version of orthodoxy, but in the original 1926 papers, he did not clearly hold to indefiniteness, and he was willing to countenance the possibility of an underlying determinism:[19]

It is natural for him who will not be satisfied with this [indeterminism of scattered states] to remain unconverted and to assume that there are other parameters, not given in the theory, that determine the individual event.[20]

Nonetheless, even then, Born was 'inclined'[21] to an indeterministic interpretation, which, he claimed, differed from both Schrödinger's and Heisenberg's interpretations, and which he described as follows:

The guiding field, represented by a scalar function ψ of the coordinates of all the particles involved and the time, propagates in accordance with Schrödinger's differential equation. Momentum and energy, however, are transferred in the same way as if corpuscles (electrons) actually moved. The paths of these corpuscles are determined only to the extent that the laws of energy and momentum restrict them; otherwise, only a probability for a certain path is found, determined by the values of the ψ function.[22]

This passage describes an interpretation that appears to have little in common with what we today call 'the Born interpretation'.

Born's statement of his interpretation is suggestive of two distinct features in the evolution of a quantum system. First, there is the wave function, which for Born is a 'guiding field'. Second, there are 'corpuscles' that follow (continuous?) paths. (This second feature is less clearly a part of Born's interpretation, because of the qualifier 'as if'. However, if there are no 'corpuscles' in Born's interpretation then what does the 'guiding field' guide?)

Although the analogy can be pressed too far, there is a similarity to Bohm's theory, in which the wave function is also taken to be a kind of 'guiding field', while particles follow continuous trajectories. I shall leave an account of Bohm's theory to chapter 5, however. Here I want only to draw the lesson that, in 1926, it was apparently *not* part of quantum-mechanical dogma that a guiding field and particles with trajectories ('paths') were out of the question.[23] In more general terms: Born apparently did not rule out the reasonableness of unorthodox interpretations.

A crucial difference between Born's 1926 interpretation and Bohm's theory is that while Bohm's theory is deterministic, Born appears to prefer indeterminism.[24] However, here too Born allows the possibility that quantum indeterminism could be underpinned by a more fundamental determinism. Indeed, Born claims that no purely physical argument could ever decide the issue:

I myself am inclined to give up determinism in the world of atoms. But that is a philosophical question for which physical arguments alone are not decisive'.[25]

And here too he makes a prophetic statement:

This possibility [of underpinning quantum indeterminism with determinate and determined values] would not alter anything relating to the practical indeterminacy of collision processes, since it is in fact impossible to give the values of the phases; it must in fact lead to the same formulae as the "phaseless" theory proposed here.[26]

Again, without giving Born more foresight than he had, we may note that *exactly* the situation he describes is what happens in Bohm's theory.

Others took more or less the same view. Jordan, for example, wrote in 1927:

The circumstance that quantum laws are laws of averages, and can only be applied statistically to specific elementary processes, is not a conclusive proof that the elemental laws themselves can only be put in terms of probability.[27]

And near the end of his article, which is mostly concerned with the status of determinism in quantum mechanics, Jordan can bring himself to ask (seriously): 'Does modern physics recognise any complete determinism?'[28] For Jordan, in 1927, this question was open. For us it should be open too.

1.3 Options for interpretation

In this section, I make 'orthodoxy' more precise, by making 'indefiniteness' and 'dynamical indeterminism' precise. In later chapters, I will classify various interpretations based on whether they accept or deny indefiniteness and dynamical indeterminism.

1.3.1 The eigenstate–eigenvalue link

Indefiniteness can be made precise as the acceptance of the so-called 'eigenstate–eigenvalue link'.[29] The eigenstate–eigenvalue link says that a system in the state W has the value a for the observable A if and only if $p^W(P_a^A) = 1$.

The eigenstate–eigenvalue link can be analyzed as the conjunction of two other conditions, which Fine has called the 'Rule of Law' and the 'Rule of Silence', and which we may call, more pedantically, the 'eigenstate to eigenvalue link' and the 'eigenvalue to eigenstate link'. The former is the 'if' part of the eigenstate–eigenvalue link, and the latter is the 'only if' part.

Hence the eigenstate to eigenvalue link says that if $p^W(P_a^A) = 1$, then the system (whose state is W) has the value a for the observable A. In Fine's terms, if for some eigenvalue, a, of the operator, A, the state, W, of a system is such that $p^W(P_a^A) = 1$, then the 'law' requires that we attribute the value a to the system.

The eigenvalue to eigenstate link says that if the system (whose state is W) has the value a for the observable A, then W is such that $p^W(P_a^A) = 1$. In Fine's terms, if there is *no* eigenvalue, a, of A such that $p^W(P_a^A) = 1$, then we must be silent about the observable A.

We may weaken the Rule of Silence a bit. Indeed, it would seem to need weakening. To see why, consider an operator, A, with the spectral resolution

$$A = \sum_i a_i P_{a_i}^A. \tag{1.12}$$

(Recall that by the spectral decomposition theorem, self-adjoint operators can always be written as a weighted sum of their eigenprojections, where the weights are the corresponding eigenvalues.) Now consider a second operator, $P_{a_1 \vee a_2}^A$, the projection onto the subspace spanned by two eigenspaces, $P_{a_1}^A$ and $P_{a_2}^A$, of A; $P_{a_1 \vee a_2}^A$ is self-adjoint and is therefore a genuine observable. Its only eigenspace (with non-zero eigenvalue) is $P_{a_1 \vee a_2}^A$ itself, corresponding to the eigenvalue 1. Finally, consider a statevector, $|\psi\rangle$, in the subspace spanned by $P_{a_1}^A$ and $P_{a_2}^A$. In this case, $p^\psi(P_{a_1 \vee a_2}^A) = 1$ but $p^\psi(P_{a_i}^A) \neq 1$ for all i. The Rule of

Silence then says that we may not speak about the observable A, while the Rule of Law *requires* that we attribute the value 1 to the observable $P^A_{a_1 \vee a_2}$.

However, if $P^A_{a_1 \vee a_2}$ has the value 1, then it is natural to suppose that the value of A is at least restricted to $\{a_1, a_2\}$. In saying so, we do not give up indefiniteness. The claim is not that the value of A is either a_1 or a_2, but only that its value is 'restricted to' the range $\{a_1, a_2\}$. Of course, some metaphysical work is needed to make it clear what it means to have a value restricted to a given set, while not having one of the values in the set, but any advocate of indefiniteness already has that work to do anyway. (Quantum mechanics already tells us that the value of an observable is restricted to the set of its eigenvalues. Indefiniteness adds that sometimes an observable has none of the values in this set.)

Whether we accept this weakened version of the Rule of Silence or not, the basic idea is the same: there are situations in which a system has no particular value for a given observable. Instead, we say that the system's value is 'restricted to' (some subset of) the possible values for the observable in question. In this sense, the system is in an indefinite state, relative to the observable in question. Once again, we may distinguish this doctrine from an epistemic interpretation of quantum probabilities, according to which it may be that a system in the state $|\psi\rangle$ has the value a for A while nonetheless $p^\psi(P^A_a) \neq 1$.

1.3.2 Determinism and indeterminism

Earlier in this chapter, I defined 'dynamical indeterminism' in terms of the results of a measurement. However, the concept is clearly more general, and it is best to have a definition of determinism that does not rely on the notion of measurement. Hence let us say that an interpretation is dynamically indeterministic if it asserts that the state of a system at a time t cannot in general be predicted with certainty given the history of its states prior to t. When no ambiguity results, I shall refer to dynamically indeterministic theories as just 'indeterministic'.

Although it will not be necessary for the issues raised in this book to grapple with the difficult questions that surround many discussions of determinism, a few words by way of connection to existing discussions will help to make more clear what is meant by 'indeterminism' here.

There are several interpretations of 'indeterminism' as I have defined it thus far, corresponding to several accounts of what it means for a theory to predict with certainty the occurrence of some future event, given the presently occurrent events. One convenient way to see the possibilities is in the responses to the question 'How can a theory fail to permit prediction with certainty?' There are two sorts of obstacle to prediction with certainty: inability to determine future states from initial states, and impossibility of knowing initial states in the first place.

Each of these obstacles comes in two forms, an 'in practice' form, and an 'in principle' form. For example, the inability to use a theory to calculate the state of a system at time t given its states up until t may be due simply to the difficulty (for us, or for our instruments) of performing the calculation. This type of difficulty is familiar from both quantum mechanics and classical mechanics — indeed, very few instances of the Schrödinger equation have ever been solved explicitly, and it can be very difficult — in some cases, practically impossible — to guarantee that results obtained by making approximations (and applying perturbation theory) are close to the 'real' answer. Nonetheless, Schrödinger's equation is a linear differential equation, and given certain constraints, we can be assured that there *exists* a solution, even if we are unable to find it. On the other hand, there may be some more fundamental reason that a theory cannot be used to calculate the later state of a system, given its earlier states. It might be, for example, that the states of some chaotic systems resist prediction arbitrarily into the future by *any* predictive instrument. For example, the complexity of the calculation might grow exponentially as the distance into the future increases.

Similarly, it might be that we cannot make predictions with certainty because we cannot know enough about a system's history (or, its present state) to make the prediction. This situation too is familiar from both classical mechanics and quantum mechanics. Indeed, it is arguable that we will never be in a position to know exactly the state of any given system, because of the *de facto* limitations on our ability to gather information. On the other hand, it may be that there is a reason in principle that we cannot have the knowledge required. This situation occurs in Bohm's theory, as we will see in chapter 5: if Bohm's theory is true, then one *cannot*, as a matter of principle, know the complete (present) state of a system.

Theories that do not allow prediction with certainty for any of these reasons are not necessarily indeterministic, on the account being offered here. The idea that I will attempt to make precise presently is that in spite of the impossibility — in practice or in principle — of acquiring the necessary information, or of using it to make calculations, a theory might nonetheless postulate that the complete state of a system at a given time (or, the history of its states up to a given time) fixes its complete state at any later time.

The modifier 'complete' is important. If all it took for determinism were prediction with certainty of some aspect or other of a system's state, then determinism would be almost unavoidable. Consider, for example, a particle undergoing Brownian motion (a random walk) with finite velocity, v. In this case, knowing the particle's position, x, at a time t_0 does allow us to make some predictions with certainty, namely, that the particle will be within $v \times (t - t_0)$ of x at time $t > t_0$. Nonetheless, we cannot predict with certainty the exact position of the particle at time t.

Moreover, we must expand the notion of a 'system' to the entire universe. After all, it may be that a theory is completely deterministic at the level of the universe, but due to interactions, indeterministic at the level of any subsystem of the universe.

These qualifications lead naturally to a definition such as the following:[30] A theory, T, is deterministic if at every time, t, the complete state of the universe at time t (or, the history of the complete states of the universe up through time t) is consistent (according to T) with only one complete state at time t', for any $t' > t$. If T is Markovian,[31] then the parenthetical alternative is unnecessary. Also, if T is time-homogenous,[32] then it suffices to say that for all t', the initial complete state of the universe is consistent with only one complete state at $t' > 0$.

I have taken for granted that we know what it means for a theory to say that a complete state at one time is or is not consistent with a given state at a later time. There are different ways of making this idea precise, corresponding to different accounts of scientific theories. On the 'axiomatic' account (according to which a scientific theory is a set of axioms plus rules of inference), consistency of some initial state, S_1, with some later state, S_2, amounts to the *logical* consistency of the proposition 'the system has S_1 initially and S_2 later' with the axioms, in the context of the rules of logical inference. On the 'semantic' account (according to which a scientific theory is a set of models, or histories of the universe), consistency of S_1 and S_2 amounts to there being some model in which S_1 is the state at the initial time, and S_2 the state at the later time.

When applying these definitions, one must be careful to distinguish actual states of the universe from probability measures over them. Indeed, if we take the statevector for the universe to be its *state* in the present discussion, then quantum mechanics turns out to be deterministic (whether we adopt the syntactic or semantic conception of theories). However, this fact about quantum-mechanical statevectors is of little interest in this discussion, unless we have *also* adopted the principle that the statevector for the universe is also the *complete state* of the universe — in other words, unless we adopt the eigenstate–eigenvalue link. As soon as we admit that the statevector for the universe is a probability measure over possible states of the universe, then the question of determinism is open.

1.3.3 The lay of the land

Given the options of the previous two sections, we may classify interpretations of quantum mechanics as being of one of four types. Each type faces problems.

First, there are orthodox interpretations, which accept the eigenstate–eigenvalue link and indeterminism. These interpretations immediately face the measurement problem: given the eigenstate–eigenvalue link, quantum mechanics apparently ascribes the wrong state to systems at the end of a measurement (and probably other times as well).

Second, there are interpretations that accept the eigenstate–eigenvalue link, but are deterministic. These theories seem to face two problems. Because they accept the eigenstate–eigenvalue link, they face the measurement problem. Also, they must explain why quantum mechanics, as we use it at least, is a probabilistic theory. If the true theory of the world is *really* deterministic, then from where do quantum probabilities come?

Third, there are interpretations that deny the eigenstate–eigenvalue link and are indeterministic. These interpretations have a built-in way to solve the measurement problem (though they must still convince us that they do so — denying the eigenstate-eigenvalue link is not *sufficient* for solving the mesaurement problem), but they face their own difficulties. In particular, they must have a convincing way to avoid the Dutch Book, for by denying the eigenstate–eigenvalue link, they apparently adopt some version of an epistemic interpretation of quantum-mechanical probabilities.

Fourth, there are interpretations that deny the eigenstate–eigenvalue link and are deterministic. Like interpretations of the previous category, these interpretations can exploit their denial of the eigenstate–eigenvalue link to help solve the measurement problem, and like them, these interpretations must explain how they avoid the Dutch Book. In addition, they face the same problem as interpretations of the second category, namely, explaining why we have thus far been able to make only probabilistic predictions with quantum theory.

In the next four chapters, I will consider at least one interpretation in each of these categories. By way of warning (and excuse): my consideration of these interpretations will not be complete. Indeed, a complete assessment of any one of them would require a book in itself. It is also not my aim to evaluate these interpretations (though I shall not shy away from stating my opinions). Rather, my aim is to *use* these interpretations in the investigation of probability and non-locality in quantum mechanics.

2

Orthodox theories

2.1 How is orthodoxy possible?

How can an interpretation maintain both the eigenstate–eigenvalue link *and* indeterminism? Given the former, the properties possessed by a system are completely fixed by its quantum-mechanical state, but the quantum-mechanical state evolves deterministically, as I noted at the end of chapter 1. By themselves, then, the eigenstate–eigenvalue link and the quantum-mechanical equation of motion lead to determinism. Orthodoxy must change one of these things if it wants to maintain indeterminism.

Of course, it cannot change the eigenstate–eigenvalue link, lest it no longer be orthodoxy. Hence it changes the equation of motion. In this chapter, I will discuss two ways to change the quantum-mechanical equation of motion: by 'interrupting' it from time to time with some other (indeterminsitic) equation, or by making a wholesale replacement. The first strategy I discuss in the next section, and the second in the subsequent section.

2.2 The projection postulate

2.2.1 Collapse as an analogue of Lüder's rule

Thus far, we have been working in the 'Schrödinger picture', according to which states evolve in time (according to the Scrödinger equation) and any given observable is at all times represented by the same operator. The Heisenberg picture reverses things: the states are constant in time and the operators representing observables change. If $A(0)$ represents a given observable at time 0, then, in the Heisenberg picture, $A(t)$ represents the same observable at time t, with

$$A(t) = U^{-1}(t)A(0)U(t). \tag{2.1}$$

The Schrödinger and Heisenberg pictures are said to be equivalent, in the sense that they generate the same probability measures over the values of observables at all times. In the Heisenberg picture, the probability rule (1.3) is, in terms of the values of observables:[1]

$$p^W (A(t) \text{ takes value } a) = \text{Tr}[W P_a^{A(t)}]$$
$$= \text{Tr}[W P_a^{U^{-1}(t)AU(t)}]$$
$$= \text{Tr}[W U^{-1}(t) P_a^A U(t)], \qquad (2.2)$$

where $A = A(0)$. However, the trace functional is invariant under cyclic permutations of its arguments. Hence

$$\text{Tr}[W U^{-1}(t) P_a^A U(t)] = \text{Tr}[U(t) W U^{-1}(t) P_a^A]. \qquad (2.3)$$

Now note that $U(t) W U^{-1}(t)$ is the state of the system at time t in the Schrödinger picture. Hence, (2.2) is equivalent to (1.3). The Schrödinger and Heisenberg pictures are, for this reason, said to be predictively equivalent.

Now recall Lüder's rule from chapter 1, which said:

$$p^W (P'|P) = \frac{\text{Tr}[P W P P']}{\text{Tr}[W P]}.$$

'Translating' Lüder's rule to the Heisenberg picture yields an expression for the conditional probability of $P'(t')$ given $P(t)$ $(t' > t)$:

$$p^W \left(P'(t')|P(t) \right) = \frac{\text{Tr}[P(t) W P(t) P'(t')]}{\text{Tr}[W P(t)]}. \qquad (2.4)$$

Now let $t = 0$, and substitute $U^{-1}(t') P'(0) U(t')$ for $P'(t')$:

$$p^W \left(P'(t')|P(0) \right) = \frac{\text{Tr}[P(0) W P(0) U^{-1}(t') P'(0) U(t')]}{\text{Tr}[W P(0)]} \qquad (2.5)$$

then, using the invariance of the trace functional under cyclic permutations of its arguments, we get

$$p^W \left(P'(t')|P(0) \right) = \frac{\text{Tr}[U(t') P(0) W P(0) U^{-1}(t') P'(0)]}{\text{Tr}[W P(0)]}. \qquad (2.6)$$

Equation (2.6) is very suggestive. Translating back to the Schrödinger picture, it suggests that the occurrence of P at $t = 0$ *changes* the state of the system from $W(0)$ to $P(0) W P(0) / \text{Tr}[W P(0)]$, which then evolves as usual according to the unitary operator, $U(t')$. This evolved state is then used to calculate the probability of P' at the time t'. In other words, the occurrence of P at time 0 'collapses' the state W, or 'projects' it onto P (and renormalizes).

However, we can already see that this argument for collapse is flawed. The flaw is in the move from Lüder's rule to (2.4). This move must involve some hidden assumption, because Lüder's rule does not involve time-evolution at all, while (2.4) does (except, of course, when $t = t'$, in which case (2.4) is just Lüder's rule).

Nonetheless, (2.4) does look to be in the *form* of Lüder's rule, and although it is really a kind of *transition* probability, transition probabilities *are* conditional probabilities of a kind. Why, then, can we not 'derive' (2.4) just as one derives Lüder's rule? Recall from chapter 1 that we can get Lüder's rule from the condition that $p^W(P'|P) = p^W(P')/p^W(P)$ when $P' \subseteq P$. (See (1.5).) The formal analogue of this condition in the present case is that

$$p^W\left(P'(t')\big|P(t)\right) = \frac{p^W\left(P'(t')\right)}{p^W\left(P(t)\right)} \tag{2.7}$$

when $P'(t') \subseteq P(t)$ and $t' > t$. Starting from this condition, one can easily follow the steps of the derivation of Lüder's rule to derive (2.4).

Is (2.7) at all plausible? To answer this question, let us ask first why the condition (1.5) is plausible.

The main idea is that the occurrence of P does not change the *relative* probabilities of events 'contained in' P (i.e., events whose occurrence implies the occurrence of P). The most straightforward justification for this assumption is based on an ignorance interpretation of the measure p^W. Having learned that P is occurrent, we can say that it is occurrent *because* some $P' \subset P$ is occurrent. However, because the occurrence of *any* $P' \subseteq P$ is guaranteed to make P occurrent, the occurrence of P does nothing to revise the relative probabilities of the $P' \subseteq P$.

We may make the point slightly more graphically as follows. Consider an ensemble of systems, all in the state W. Restricting to the subensemble of systems for which P occurs does not change the ratio of of the number of systems for which P' occurs to the number of systems for which P'' occurs, if $P', P'' \subseteq P$. For *every* system for which any $P' \subseteq P$ occurs is in the subensemble for which P occurs. Hence we have

$$\frac{p^W(P'|P)}{p^W(P''|P)} = \frac{p^W(P')}{p^W(P'')}. \tag{2.8}$$

Letting $P'' = P$ yields (1.5).

Of course, orthodox interpretations cannot use this motivation for Lüder's Rule, because they do not adopt the ignorance interpretation of p^W. The closest they can come, perhaps, is to say that the occurrence of P should not change the *shape* of the probability measure, p^W, restricted to events in

P. However, this claim is just (2.8) in words, and it remains unclear why it should be true in the orthodox interpretation.

Anyhow, let us not pause to consider whether Lüder's rule is well motivated in orthodox theories. The main point here is that the orthodox interpretation has no good motivation for (2.7). It therefore cannot motivate (2.4) in the same way that one can motivate Lüder's rule via (1.5).

Indeed, (2.7) is not even plausible classically. Imagine that the sample space (in a classical probability theory) is $\Omega = \{\omega_1, \omega_2, \omega_3\}$ and let $P = \{\omega_1, \omega_2\}$. Consider a probability measure over Ω given by $p(\omega_i) = 1/3$ for $i = 1, 2, 3$. For convenience, suppose that p is constant over time. Certainly the following (deterministic) probabilities are not *a priori* unreasonable:

$$p(\omega_1 \text{ at } t' | \omega_3 \text{ at } t) = 1,$$
$$p(\omega_2 \text{ at } t' | \omega_1 \text{ at } t) = 1,$$
$$p(\omega_3 \text{ at } t' | \omega_2 \text{ at } t) = 1.$$

However, (2.7) forbids these transition probabilities. According to (2.7), $p(\omega_1|P) = 1/2$, while the transition probabilities above entail $p(\omega_1|P) = 0$.

More generally, the point is that (2.7) rules out transition probabilities that are allowed by the single-time probabilities at t and t'. The corresponding problem does not arise for (1.5) because it is a single-time probability.

However, there is at least one legitimate way to motivate (2.7), and it relies on the following two principles:

(i) A 'strong' ignorance interpretation of p^W: probability measures must always be revised in the light of new knowledge.

(ii) Over periods where our knowledge (about occurrent events) is otherwise unchanged, probability measures evolve according to the usual equations of motion of quantum mechanics.

The first principle suggests once you learn that P, you revise the probability measure p^W from $\text{Tr}[W \cdot]$ to $\text{Tr}[PWP \cdot]/\text{Tr}[WP]$. This probability measure is generated in the usual way by a quantum-mechanical state, $\widetilde{W} = PWP/\text{Tr}[WP]$. Applying principle (ii) to the measure $p^{\widetilde{W}}$, we find immediately that (2.4) holds, because the evolution of $p^{\widetilde{W}}$ is generated by the evolution of $\widetilde{W}$: $\widetilde{W}(t) = U(t)\widetilde{W}(0)U^{-1}(t)$.

Of course, this derivation of (2.4) is unavailable to orthodox interpretations, because, again, it relies on the ignorance interpretation of p^W. Indeed, it relies on a rather strong form of the ignorance interpretation — one that is basically tantamount to the projection postulate itself. It seems that, in the end, the principles of orthodoxy do not entail, or even plausibly motivate, equation (2.4).

2.2.2 The projection postulate and its problems

Orthodoxy must, therefore, accept (2.6) as a postulate, and one that does not clearly sit well with the principles of orthodoxy. There is a further problem: (2.6) is itself ambiguous. It tells us to 'collapse' the state when an event, P, 'occurs', but when *does* an event 'occur'? For (2.6) to be a well-defined prescription, it must specify what counts as 'occurrence' of an event. Doing so leads to the usual formulation of the projection postulate:

Projection postulate: Upon measurement of an observable, A, on a system, σ, in the state, W, the state of σ 'collapses' to $P_a^A W P_a^A / \mathrm{Tr}[W P_a^A]$ for some eigenvalue, a, of A. The probability that the state collapses to $P_a^A W P_a^A / \mathrm{Tr}[W P_a^A]$ is $\mathrm{Tr}[W P_a^A]$.

When stated as above, the projection postulate is extremely suspect, if not grossly *ad hoc*.[2] It is explicitly designed to solve the measurement problem. Moreover, supposing that quantum mechanics is to be considered a fundamental theory — the theory of quantum phenomena, from which we are to derive our theories of everyday phenomena — it is extremely unsatisfactory to include 'measurement' as a primitive notion in the theory. At the least, we should require that the notion of 'measurement' be unambiguously reducible to fundamental elements (in particular, processes, or interactions) of the theory. Thus far, no fully satisfactory reduction has been proposed.

However, although nobody has (yet?) shown how to eliminate 'measurement' from the statement of the projection postulate, it may be that there are good reasons nonetheless to suppose that collapse does occur upon measurement. Of course, I am not now asking why collapse occurs at all, but why it should occur upon measurement.

One answer is suggested by von Neumann's discussion of repeated measurements.[3] His discussion suggests that in the case of repeated measurement, collapse is required to get the transition probabilities right. The empirical phenomenon that must be recovered is that when we repeat a measurement, the result of the second measurement always matches the result of the first, assuming that the system does not evolve between the measurements.[4]

The projection postulate is certainly sufficient to guarantee this matching, because the state of the system after the first measurement will lie in the subspace P_a^A, so that the probability of P_a^A at the later time is 1; but is the projection postulate necessary for this result? Apparently not. Indeed, if we examine the phenomenon of repeated measurement more closely, then it seems that nothing more than quantum probability theory and the Schrödinger equation are required to guarantee matching. Consider again what happens during a typical (but ideal!) measurement-interaction, supposing that we are to measure the observable represented by A and that the system begins in the state $\sum_i c_i|a_i\rangle$, where the $|a_i\rangle$ are eigenvectors of A. The measurement-interaction induces the evolution:

$$\sum_i c_i|a_i\rangle|M_0\rangle \quad \longrightarrow \quad \sum_i c_i|a_i\rangle|M_i\rangle,$$

where, as in chapter 1, $|M_i\rangle$ are states of the measuring apparatus. Now introduce a second apparatus, whose states are $|N_i\rangle$. The second measurement-interaction gives

$$\sum_i c_i|a_i\rangle|M_i\rangle|N_0\rangle \quad \longrightarrow \quad \sum_i c_i|a_i\rangle|M_i\rangle|N_i\rangle.$$

Given the final state, the probability of finding the apparatuses in non-matching states is zero:

$$p^\Psi(P_j^M \otimes P_k^N) = \left|\left(\sum_i c_i\langle a_i|\langle M_i|\langle N_i|\right)P_j^M \otimes P_k^N\left(\sum_i c_i|a_i\rangle|M_i\rangle|N_i\rangle\right)\right|^2$$
$$= \delta_{jk}|c_j|^2,$$

where $\delta_{ij} = 0$ if $i \neq j$ and $\delta_{ij} = 1$ if $i = j$.

So why would anybody think that the projection postulate is needed to guarantee matching of results for repeated measurements? Perhaps because the account just given does not seem *explanatory*. Although it does guarantee that the apparatuses will match, it may not be a sufficient explanation for *why* they match. On the other hand, if the *state* of the system after the first measurement is $P_a^A W P_a^A/\text{Tr}[W P_a^A]$, where P_a^A is the result indicated by the apparatus, then somehow we seem to have an explanation for why the result of the second measurement is P_a^A.

Or so the argument goes. The initial plausiblity of this argument rests on the presumption that the only explanation for the coincidence between the apparatuses is that they measured some system that was in the 'same' state for each measurement. However, if we adopt the projection postulate, then they did no such thing. The first apparatus measured a system in the state $\sum_i c_i|a_i\rangle$, while the second measured a system in the state $|a_n\rangle$ for some n.

Perhaps, then, the argument is that the agreement of the second apparatus with the first must be explained by its *having* to agree. That is, by collapsing the system onto $|a_n\rangle$ (for some n) after the first measurement, the probability of disagreement is zero — the second apparatus must show the result $|a_n\rangle$. However, this argument also fails, for the probability of disagreement is *already* zero without the projection postulate, as we saw. Collapse seems no better able to produce an explanation of the agreement than does standard quantum mechanics without the projection postulate.

So we are still left wondering why collapse should happen upon measurement. One quite natural reply is to modify the projection postulate so that collapse occurs not upon measurement, but upon 'observation' by a sentient (or rational, or human) being. The justification for this move is apparently a kind of minimalism: collapse *must* occur at least under *these* circumstances, in order to recover the phenomena of everyday experience. When an experimenter observes a measuring apparatus, the apparatus is always seen to be in a definite state. Hence, this view goes, we say that collapse occurs upon (and only upon) observation, and we justify this (modified) projection postulate on the simple basis that without it, the theory makes false predictions.

As we shall see in the next chapter, it is not clear whether the eigenstate–eigenvalue link forces one to apply collapse even here — when a system is observed — but supposing that it does, one must begin to wonder whether it is not time to apply *modus tollens*. There are two obvious problems with orthodoxy as we now have it.

The first is the problem of its being blatantly *ad hoc*. The point is not that the orthodox interpretation must adjust itself in light of empirical fact — any good theory will do so. The point is rather that the adjustments themselves do not sit well with one of the basic principles of the theory, namely, Scrödinger's equation. The orthodox interpretation must say that measurements, or observations, are somehow 'special', not described by Schrödinger's equation, but the theory resists any attempt to say precisely *what* counts as a measurement, or an observation (much less a sentient being). The theory of quantum mechanics, prior to any interpretation, makes no distinction between measurements and non-measurements, nor does there appear to be any way to make such a distinction using just the mathematical and physical principles that make up quantum mechanics.

The second outstanding problem with orthodoxy is the physical implausibility of the projection postulate. Given its adherence to the eigenstate–eigenvalue link, orthodoxy must admit that collapse is a discontinuous change in the physical state of a system. What is it about *measurement*, or *observation* (as opposed to other physical processes), that causes such a collapse? The problems with the projection postulate are serious, and do not show much hope of being solved in a satisfactory way.

In the previous section, I examined the process of collapse, asking whether it can be derived or motivated from the principles of of orthodoxy. Apparently, it cannot. Even worse, the most natural derivations of or motivations for collapse are based on principles that orthodoxy denies. In other words, the best answer that orthodoxy can give to the question 'why does collapse occur?' is 'it just does'. In this section, I allowed that collapse *occurs*, and focused on a second question, namely, 'why does collapse occur *upon measurement?*' Again, we found little motivation for the supposition that collapse occurs upon measurement.

Hence there are *two* senses in which the projection postulate is a postulate. First, the fact that collapse occurs at all is postulated rather than derived. Second, the fact that collapse occurs upon measurement (and only upon measurement) is postulated rather than derived. Moreover, as I suggested here and will discuss in more detail in the next chapter, it may be that collapse is not required for empirical success even *with* the eigenstate–eigenvalue link. These shortcomings of orthodoxy should prompt us to consider alternatives.

2.3 The continuous spontaneous localization (CSL) theory

2.3.1 Intuitive introduction to CSL

One obvious thing for an orthodox interpreter to do is to try to achieve collapse in a way that does not make essential reference to the notion of a 'measurement'. Such a program has been in existence since the 1960s, but has blossomed only in the past several years. The idea is to rewrite the Schrödinger equation so that collapse is automatically incorporated into the evolution of a system. Collapse in this case need not be tacked onto the theory, as an occasional interruption of the basic equation of motion. Instead, collapse is a consequence of the basic equation of motion.

I will focus on one realization of this idea, CSL. According to CSL, collapse of the wave function is a real physical process.[5] It is continuous in the sense that the evolution of the wave function is continuous — it is governed by a differential equation. It is spontaneous in two senses. First, the wave function at the beginning of a collapse does not determine the wave function at the end of a collapse. Second, there is no deterministic story to be told about how collapse occurs — the basic physical processes in CSL are stochastic — the evolution of the wave function is governed by a *stochastic* differential equation.

The intuitive idea behind CSL is that the wave function does not evolve deterministically, according to the Schrödinger equation. Instead, the wave function evolves according to a stochastic process that looks much like the deterministic Schrödinger evolution, with the exception of a small stochastic perturbation. Because of the way it is defined, this perturbation has a negligible effect on small systems, but generates a collapse of the wave function for large systems, under the right condition.

The 'right condition' is that the state of the large system describes a superposition of macroscopically distinct states, which are states describing a large number of particles in a superposition of states well separated in position (one lump here, one lump there). In other words, according to CSL, objects like cats and pointers are (almost) always well localized; and well localized is, after all, just how we always find them.

However, note that CSL does not quite stipulate by fiat the conditions under which collapse will occur. Instead, CSL modifies the Schrödinger equation in such a way that collapse will, as a matter of fact, almost always occur under the condition described, and in accordance with the probabilities given by quantum mechanics. The details are a bit complex, and are given in the next section, but the basic idea is simple. There is a small stochastic element added to the evolution of every particle. For a single particle, the effect of the stochastic modification is negligible. However, it is additive, so that when a system has a great many particles, the stochastic part of the evolution swamps the Schrödinger part. Moreover, the stochastic part is designed in such a way that it tends to localize the system.

By giving this account of collapse, CSL avoids the problems of the standard interpretation. Collapse is no longer triggered by measurements, which is good because it is hard to say what counts as a 'measurement' in the first place. Instead, collapse is a fundamental process affecting all systems. Indeed, because it affects large systems much more strongly than small ('quantum') systems, the CSL-collapse is also able to account for the well-localized states of large objects that are *not* involved in a measurement. On the other hand, there is no denying that the stochastic element was designed for much the same reason as the projection postulate, namely, to solve the measurement problem. The best that can be said for it, I think, is that it does (apparently) solve the measurement problem, and without making essential reference to the notion of 'measurement'. CSL is at least a demonstration that orthodoxy can solve the measurement problem by using a fundamental equation of motion instead of an *ad hoc* postulate such as the projection postulate.

2.3.2 CSL as a modification of the Schrödinger equation

Recall that evolution of a quantum system is given in terms of a unitary operator, $U(t)$, which, recall, is given by $U(t) = e^{-iHt}$. In CSL, $U(t)$ is replaced by a non-unitary operator, $\tilde{U}(t)$,

$$\tilde{U}(t) = \exp[-iHt] \exp\left[-\frac{\gamma}{2}t \int N^2(\mathbf{x})d^3\mathbf{x}\right] \exp\left[\int N(\mathbf{x})B(\mathbf{x},t)d^3\mathbf{x}\right]. \quad (2.9)$$

These extra terms need explanation. The $N(\mathbf{x})$ are a (continuous) family of 'number density' operators, defined by

$$N(\mathbf{x}) = (\alpha/2\pi)^{3/2} \int \exp\left[-\frac{\alpha}{2}(\mathbf{y}-\mathbf{x})^2\right] a^\dagger(\mathbf{y})a(\mathbf{y})d^3\mathbf{y}, \quad (2.10)$$

where $a^\dagger(\mathbf{y})$ and $a(\mathbf{y})$ are improper 'creation at $\mathbf{y}$' and 'annihilation at $\mathbf{y}$' operators (more properly: operator-valued distributions). Intuitively, $N(\mathbf{x})$ indicates roughly the number of particles inside the ball of radius $\alpha^{-3/2}$ centered at $\mathbf{x}$. The $B(\mathbf{x},t)$ is a field of Brownian motion (i.e., a continuous family of Brownian motion processes, each occurring at a point $\mathbf{x}$).[6] For reasons that will become clear in the next subsection, I call the term

$$\exp\left[-\frac{\gamma}{2}t \int N^2(\mathbf{x})d^3\mathbf{x}\right] \quad (2.11)$$

the 'decay term' (where γ is a decay constant whose value is discussed later) and the term

$$\exp\left[\int N(\mathbf{x})B(\mathbf{x},t)d^3\mathbf{x}\right] \quad (2.12)$$

the 'growth term'.

The only other mathematical ingredient in CSL is a prescription for the probabilities defining the Brownian motion processes $B(\mathbf{x},t)$. First, they are conditioned so that

$$B(\mathbf{x},0) = 0, \quad \overline{dB(\mathbf{x},t)} = 0, \quad \overline{dB(\mathbf{x},t)dB(\mathbf{x}',t)} = \gamma\delta(\mathbf{x}-\mathbf{x}')dt \quad (2.13)$$

(where an overline indicates an expectation value). Second, these Brownian processes are not governed by the usual probabilities for sample paths, $b(\mathbf{x},t)$, of $B(\mathbf{x},t)$. If we denote that measure by p_{raw}, then the ('cooked') probability for the realization of a given sample path, $b(\mathbf{x},t)$ (which is really a continuous family of sample paths, one for each value of $\mathbf{x}$), up through time t for a system with the initial wave function $|\psi\rangle$ is

$$p_{\text{cooked}}[b(\mathbf{x},t)] = p_{\text{raw}}[b(\mathbf{x},t)] \times \text{Tr}[\tilde{U}_b(t)|\psi\rangle\langle\psi|\tilde{U}_b^{-1}(t)], \quad (2.14)$$

where $\tilde{U}_b(t)$ is $\tilde{U}(t)$ as given by the sample path $b(\mathbf{x},t)$. (That is, substitute $b(\mathbf{x},t)$ for $B(\mathbf{x},t)$ in the expression for $\tilde{U}(t)$, which then becomes deterministic.)[7]

2.3.3 *Physical clarification of CSL*

The physical meaning of CSL is easiest to see when we let vectors, rather than density operators, be the states of systems.[8] In fact, the discussion will be easiest if we restrict attention to wave functions, $\psi(x)$. The probability prescription (2.14) then implies that very likely the actual sample path, $b(\mathbf{x}, t)$, of $B(\mathbf{x}, t)$ will be nearly zero for all $\mathbf{x}$ except in some region where its being large would imply that $|\psi(\mathbf{x})|^2$ is large. The growth term therefore causes $\psi(\mathbf{x}, t)$ to grow over time in some region where it is appreciable. The amount of growth depends on the value of $N(\mathbf{x})$ in the region of growth. The decay term causes uniform exponential decay at a rate $(\gamma/2) \int N^2(\mathbf{x}) d^3\mathbf{x}$. Therefore, γ may be chosen to be very small, so that only for systems with large numbers of particles will the decay rate be significant (because of the integration of $N^2(\mathbf{x})$ over space). CSL chooses γ roughly equal to 10^{-30} cm^3 s^{-1}.

The discussion above shows at an intuitive level that reduction will occur. It can be shown,[9] furthermore, that the reduced state converges to one of the (improper) common eigenvectors of the $N(\mathbf{x})$. However, will the quantum-mechanical probabilities be reproduced? Yes. The reason is that the evolution of $|\psi(\mathbf{x})|^2$, which may itself be considered a real-valued stochastic process, obeys the martingale property.[10] With that fact in mind, and assuming for simplicity that the index $\mathbf{x}$ is discrete, let $|\psi\rangle = \sum_i c_i(0)|n_i\rangle$ (where the $|n_i\rangle$ are the common eigenvectors of the N_i — the discrete analogue of the $N(\mathbf{x})$). Because $|\psi\rangle$ is reduced (nearly, in finite time) onto a common eigenvector of the N_i, for a time t after reduction we have:

$$\overline{|c_i(t)|^2} = 0 \times \Pr(c_i \to 0) + 1 \times \Pr(c_i \to 1). \tag{2.15}$$

By the martingale property, $\overline{|c_i(t)|^2} = \overline{|c_i(0)|^2}$ and, because we assumed $|\psi\rangle = \sum_i c_i(0)|n_i\rangle$, $\overline{|c_i(0)|^2} = |c_i(0)|^2$. Hence $\Pr(c_i \to 1) = |c_i(0)|^2$, as in standard quantum mechanics.

I have been saying that reduction occurs nearly onto a common (improper) eigenvector of the $N(\mathbf{x})$. The point of the 'nearly' is that reduction in CSL is not *completed* in finite time. (Hence, strictly speaking, there is no 'time, t, after reduction'.) The CSL state is, however, extremely close to one of the common eigenvectors of the $N(\mathbf{x})$ in a very short time, and the argument of the previous paragraph still goes through, with the appropriate epsilons in place. That is, the probability of a reduction (in finite time) to a state that is very close to one of the common eigenvectors of the $N(\mathbf{x})$ is very close to the quantum-mechanical probability.

Let us now drop the assumption that the index on the $N(\mathbf{x})$ is discrete, and see in more detail what is going on. The common improper eigenvectors of the operators $N(\mathbf{x})$ may be written

$$|\mathbf{z}_1, \ldots, \mathbf{z}_n\rangle = \mathcal{N} a^\dagger(\mathbf{z}_1) \cdots a^\dagger(\mathbf{z}_n)|0\rangle, \tag{2.16}$$

where $|0\rangle$ is the vacuum state and $\mathcal{N}$ is a normalization factor. The eigenvalue of $|\mathbf{z}_1, \ldots, \mathbf{z}_n\rangle$ with regard to an operator $N(\mathbf{x})$ is

$$n(\mathbf{x}) = \sum_{j=1}^{n} (\alpha/2\pi)^{3/2} \exp\left[-\frac{\alpha}{2}(\mathbf{z}_j - \mathbf{x})^2\right]. \tag{2.17}$$

Each $|\mathbf{z}_1, \ldots, \mathbf{z}_n\rangle$ describes n particles at the points $\mathbf{z}_1, \ldots, \mathbf{z}_n$. The eigenvalue $n(\mathbf{x})$ represents roughly the density of particles inside the localization volume around $\mathbf{x}$. Clearly $N(\mathbf{x})$ operating on some vector $|\mathbf{z}_1, \ldots, \mathbf{z}_n\rangle$ will give a large value only when it is 'localized', i.e., when the points $\mathbf{z}_1, \ldots, \mathbf{z}_n$ are contained within the localization region around $\mathbf{x}$. Hence ψ's being large in some region means, more precisely, that the coefficient of one of the localized basis states, obtained by expanding ψ in the $|\mathbf{z}_1, \ldots, \mathbf{z}_n\rangle$-basis, is large. (Even more precisely, it means that for some small region, Δ, the integral of $|\psi|$ written in the $|\mathbf{z}_1, \ldots, \mathbf{z}_n\rangle$-basis over Δ is large.)

Note now that we need to refine slightly the earlier statement that reduction occurs onto common eigenvectors of the operators $N(\mathbf{x})$. Some of those common eigenvectors, the non-localized ones, have very small eigenvalues for every $N(\mathbf{x})$, and therefore reduction onto these eigenvectors is practically impossible. We should say, then, that reduction occurs onto the common localized eigenvectors of $N(\mathbf{x})$, i.e., eigenvectors for which a large number of particles is contained within the localization width around some point, $\mathbf{z}$.

To see how reduction occurs, suppose that $|\pi\rangle$ and $|\pi'\rangle$ describe a pointer, a large number (10^{23}) of particles localized in regions around the points $\mathbf{z}$ and $\mathbf{z}'$, respectively. Let $|\psi\rangle = \frac{1}{\sqrt{2}}(|\pi\rangle + |\pi'\rangle)$. Now consider the action of the growth term on the first component of $|\psi\rangle$. If the actual sample path, $b(\mathbf{x}, t)$, remains close to zero in the region around $\mathbf{z}$, then the growth term will be small and the coefficient of $|\pi\rangle$ will be subject to the dominant action of the decay term. This result holds even when $b(\mathbf{x}, t)$ becomes large somewhere outside the localization region around $\mathbf{z}$, because in that case it will be multiplied almost to zero by the small tail of the Gaussian generated by the action of $N(\mathbf{x})$ on $|\pi\rangle$. However, if $b(\mathbf{x}, t)$ becomes large within the region around $\mathbf{z}$, then $N(\mathbf{x})$ and $B(\mathbf{x}, t)$ conspire to make the growth term large, and the coefficient of $|\pi\rangle$ becomes large. The action on $|\pi'\rangle$ is strictly analogous.

Now, if $\mathbf{z}$ and $\mathbf{z}'$ are close together ($\ll \alpha^{-1/2}$ apart), then the reduction process will be unable to distinguish between the components of $|\psi\rangle$ and

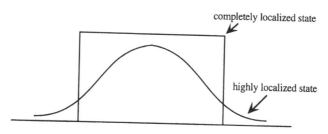

Fig. 2.1. *A completely localized state and a highly localized state.* A system in the completely localized state can be said to be confined to the region $[-x, x]$, while a system in the highly localized state is not confined to $[-x, x]$, though 'most' of its wave function is in this region.

no reduction will occur. However, if $|z - z'| \gg \alpha^{-1/2}$, i.e., if the two pointer readings are macroscopically distinguishable, then reduction to one component will occur, because the probability prescription (2.14) virtually ensures that $b(\mathbf{x}, t)$ will be large in the localization region around z or z', but not both. Hence one component of $|\psi\rangle$ decays and the other grows. With a judicious choice for γ, reduction occurs faster than can be detected.

2.3.4 Does CSL describe our experience?

It is the pronounced aim of CSL to eliminate the possibility of superpositions of macroscopically distinct states as observed measurement-outcomes, and it is the claim of CSL that reduction to states that are highly localized (as given by the $N(\mathbf{x})$ defined in (2.10)) is sufficient to achieve this aim. Two questions immediately arise, which I address in the next two subsections.

First, 'highly localized' is not 'completely localized'. For readers who might have had trouble with the previous two sections, the difference is illustrated in figure 2.1. There, the difference is illustrated for a one-dimensional wave function, but the move to many-particle wave functions (in a space of $3N$ dimensions — three for each particle) does not bring any relevant new features. CSL never effects a reduction to a *completely* localized state (except in infinite time). Is it sufficient that it effects reduction to merely highly localized states?

Second, accepting that 'highly localized' is localized enough, and accepting that paradigmatic measurements result in highly localized states of macroscopic objects, we may still wonder whether *all* measurement-outcomes — much less all human observations — are adequately described in this way.

2.3.4.1 Vagueness

The reduction of the wave function in CSL is never 'complete', except in the limit of infinite time. That is, the reduced states have 'tails' extending to infinity. This fact might be thought to raise a problem for the interpretation of the reduced state, for when a measurement is complete, and the pointer is at **z**, we do not say that it is at **z** 'with high probability'. We say instead that it is at **z** definitely, with probability 1. But then what *we* say and what CSL says are different, surely to the discredit of CSL.

The proponents of CSL have taken the view that, in CSL, the squared norm of the wave function is not to be interpreted as a probability, but instead as 'density of stuff'.[11] Although the details would need some work, the basic point is clear enough: the wave function represents a continuously evolving 'fluid-like' stuff, which we (who are made of the same stuff) witness only when it is sufficiently dense and sufficiently macroscopic. Hence, for example, when the pointer 'reduces' to a state, $|\pi\rangle$, that is highly localized around the point **z**, we witness the macroscopic and dense stuff around **z**, but do not witness the very rare (i.e., not dense) stuff around a distant point, **z**′, though, according to this view, it is there.

This interpretation of highly localized states raises an important point about the interpretation of quantum-mechanical 'observables' generally. Note, first, that this interpretation *obeys* the eigenstate–eigenvalue link, so that it admits that when the system is in the state $|\pi\rangle$, it is in an eigenstate of some observable *other* than the position observable. However, beyond this point, which is uncontroversial (at least for one who accepts the eigenstate–eigenvalue link), it is not obvious how to characterize what the state $|\pi\rangle$ means physically. Indeed, there is no canonical way to interpret *almost every* quantum-mechanical observable — only a very few of them have an agreed physical meaning. Although the interpretation of highly localized states in terms of 'density of stuff' might be unappealing to some, it does not appear to violate any well-established interpretation of observables that are close to, but not exactly, the position observable.

However, this interpretation is certainly not compulsory. A reduced CSL state is a highly localized state, sharply peaked around the 'right' macropscopic state. The probability (after reduction) that the system is in *this* state is 1, and therefore the only question is whether these highly localized states are the 'right' ones. That is, do they properly describe macropscopic objects as we witness them? I see no reason to claim that they do not. The only reason that presents itself is that the 'right' states must be *completely* localized, not merely highly localized. However, that condition on the 'right' states is too strong. There are good reasons to believe that any quantum-mechanical description of a macroscopic object (of the sort that we typically witness) will not be completely localized. Wave functions spread. Indeed, an infinite amount of energy is typically required to keep a wave function from spreading. Therefore, we are left with a choice: either quantum mechanics fails altogether to describe macroscopic objects, or the true state of those objects is highly, but not completely, localized. Taking the second alternative — I shall not consider the first here — leaves open the possibility that the reduced CSL states, which are highly localized, accurately describe macroscopic objects, i.e., that they are the 'right' states. It also leaves open the metaphysical interpretation of the CSL state.

The point I am making is worth putting in a slightly different way, in terms of the vagueness of our everyday talk about position. It seems reasonable to suppose that our everyday talk about, or beliefs about, the positions of things is not so precise that it needs representing with a single quantum-mechanical observable. Suppose, for example, that I say, 'I believe that the train is at position $\mathbf{x}$'. You reply, 'well, I can see why you would think so, but in fact the property that it has is not "being at the position $\mathbf{x}$", but a different property, very much *like* "being at the position $\mathbf{x}$"'. In that case, it seems that I am in no position to dispute your claim.

Of course, a lot hangs on what is meant by 'a property very much *like* "being at the position $\mathbf{x}$"'. There is room to wonder about what the 'closeness' of properties might mean, but it seems a reasonable sign of the 'closeness' of. two properties, P and P', that distinguishing them would be very difficult, and there is good reason to suppose that, for example, $|\pi\rangle$ is very difficult to distinguish from a completely localized state, because $|\pi\rangle$ and a completely localized state assign nearly the same probabilities to all events. Note, however, that I am not making the strong claim that the 'closeness' of two properties is simply *defined* to be 'statistical similarity'. Rather, whatever it means for two properties to be 'close', one feature of 'closeness' is statistical similarity. For example, there is a respectable sense in which two colors can be 'close' and, moreover, two colors that *are* very close *also* have the property of being very difficult to distinguish.

The idea, then, is that because it would be extremely difficult to do any experiment that could distinguish being highly localized around $\mathbf{z}$ from being completely localized at $\mathbf{z}$, it is plausible to suppose that these two properties are close, in the sense that our everyday talk and beliefs about positions are not determinate enough to single one of them out as what we 'really' mean.

Finally, note that I am *not* claiming that because $|\pi\rangle$ is very close to a completely localized state, we may close the gap, and consider $|\pi\rangle$ itself to be a completely localized state, or consider 'being in $|\pi\rangle$' to be the same as 'being completely localized'. Such proposals have been made,[12] but besides being obviously *ad hoc*, they have other failings. On this account, our *talk* is completely precise, but our *theory* is vague. When we say 'competely localized' we mean 'completely localized' and nothing else. When our theory says 'highly localized', however, it might really mean 'completely localized'. Then, of course, we need an account of *when* our theory means what it says, and when it does not, which would seem to require placing some threshhold on 'closeness'. Once such a threshhold is defined, 'vagueness' no longer means what it did before. While talk about position is indeterminately represented by any of a number of observables, the range of observables that can represent position is now precisely defined. There are no 'fuzzy boundaries' — but it seems to be a feature of vagueness that there *are* fuzzy boundaries. For example, talk about colors is vague, and in some cases, it is not clear whether a given color is 'red' or not.[13]

The advantage that the account in terms of 'density of stuff' has over this more general argument (that the 'completely localized' and 'highly localized' are 'close', in a sense that makes 'highly localized' acceptable) is that the account in terms of 'density of stuff' tells an honest ontological story about what it *means* for two properties to be 'close'. I have explicitly avoided doing so, leaving such accounts to others. The point here is that even if one does not like the interpretation of highly localized states in terms of 'density of stuff', there is good reason to believe that close is good enough.

2.3.4.2 Why John does not matter

Recall the second question raised earlier: are all measurement-outcomes, or human observations, described by highly localized states of macroscopic objects? Albert and Vaidman have argued 'no', and claim that this fact poses an irreconcilable problem for CSL.[14] Their argument is by construction of an example, a standard Stern–Gerlach spin-measurement in which the z-spin of a spin-1/2 particle is measured through deflection of the particle by an inhomgeneous magnetic field. The result is a flash at one of two well-separated detectors, L and R. The flash occurs when a particle strikes the detector, exciting a small number of atoms, which, upon returning to the ground state, emit a small number of photons.[15] An application of the standard Schrödinger equation (which is an adequate approximation to the CSL equation for small systems) shows that at this point in the process, the photons are in a superposition correlated to the spin of the measured particle:

$$|\psi\rangle = \frac{1}{\sqrt{2}}\Big(|z,+\rangle|L_{emit}\rangle|\neg R_{emit}\rangle + |z,-\rangle|\neg L_{emit}\rangle|R_{emit}\rangle\Big), \qquad (2.18)$$

where $|L_{emit}\rangle$ is a state in which photons are emitted at the left detector, and so on. Albert and Vaidman correctly claim that $|\psi\rangle$ as given in (2.18) does not experience a rapid CSL reduction. The number of particles is too small.

We shall see that there is an adequate response to this example, but first, it is useful to highlight two distinct features of the example, which I do by changing it slightly. Imagine a (highly idealized!) measuring device that accurately indicates the spin of a particle by flashing a light, and suppose that prior to the light's flashing there are no macroscopic parts of the device involved in the measurement. For spin-down (in some given basis) the device does not flash. For spin-up (in the same basis) it flashes at a low but discernible intensity. That is, the number of photons emitted is not macroscopic by CSL standards. If a particle prepared in a superposition of states of spin is measured by our device it will evolve into the following superposition:

$$\frac{1}{\sqrt{2}}\Big(|z,+\rangle|\text{flash}\rangle + |z,-\rangle|\text{no flash}\rangle\Big). \qquad (2.19)$$

Does CSL predict that this superposition will evolve rapidly into one state or the other? No, it does not, and now not for one reason, but for two. First, unlike Albert and Vaidman's case, in this case no macroscopic distances are involved. The distance between $|\text{flash}\rangle$ and $|\text{no flash}\rangle$ is much less than $\alpha^{-1/2}$. (Note that the 'low intensity' requirement could therefore be dropped. We could be speaking of a *macroscopic* number of particles with no reduction.) Second, as for Albert and Vaidman's case, the number of particles involved is not enough to trigger a rapid reduction. This version of their argument therefore involves truly microscopic superpositions; yet the difference between the superposed states is nonetheless directly observable. More generally, this second *gedankenexperiment* reveals, first, that some macroscopically distinguishable states are not well separated in position-space and, second, that some observable (by us) systems do not have large numbers of particles. Therefore, this *gedankenexperiment* would appear to present a terrible problem for CSL's hope to account for our experience, but in fact it does not.

Albert and Vaidman suggested that in order to resolve the question of whether CSL can accommodate their example, one must consider the physics of perception. The result of such a consideration is[16] that according to the presently accepted theory of perception, the act of perception itself involves states of the kind required for a reduction in CSL. I am speaking of the physical act of perception. The argument has nothing whatsoever to do with consciousness, but is entirely concerned with the physical processes involving the nerve cells that participate in perception. Those processes involve sufficiently large numbers of particles in sufficiently well-separated states to obtain a reduction in CSL. Consciousness is not required in this process. Nonetheless, the physical act of perception itself guarantees a CSL reduction.

However, Albert has a second argument,[17] based on a science-fiction creature named John, whose perceptual processes do not involve macroscopic distances or macroscopic numbers of particles. John has direct mental access to the quantum world, and has immediately occurrent beliefs about it. Albert concludes that John's state after one of his quantum observations must be a superposition of beliefs about the quantum world, and such a result is exactly what CSL hopes to avoid. The only way now to avoid this problem, says Albert, is to stipulate without reason that John is mistaken about his own beliefs. 'That's the fundamental trouble,' says Albert, 'the necessary trouble.' But is it necessarily trouble?

Of course, as far as we know John does not exist. Albert's argument depends, however, on John's being at *least* a possible person. Is John possible? I think he is. I can imagine a world that contains John. However, I can also imagine a world where gravity is repelling rather than attracting. The problem, of course, is that imagining this world requires me to deny the laws of physics in my world. Does the same problem apply to John? Does consciousness require a certain macroscopicity of the brain? Is John therefore banished in principle from all worlds that share our laws? If he is, then Albert's argument is irrelevant. It amounts only to the claim that in some other possible world, some world where the laws of physics and biology are different, CSL is an inadequate model. But then nobody would have doubted that it is, for those worlds. The problem of John becomes serious only when accompanied by an argument that his (physically) possible existence in our world is at least plausible. Whether John is physically possible is an empirical question, and one about which I am not prepared to speculate, but it is not at all obvious that he is.

However, even if John can exist in our world, Albert's argument lacks real force, for not having met John, and not having had his kind of experience, it is difficult to know whether his experience of quantum phenomena really would violate the predictions of CSL. Who could say that John might not have the perfectly definite belief: 'The atom has spin $(1/\sqrt{2})(|z,+\rangle + |z,-\rangle)$'?

In that case, there is no need to suppose John to be mistaken about his own beliefs. Or maybe, and more radically, John is capable of forming superposition-beliefs, what we might call 'non-definite' beliefs. In that case again, we need not suppose that John is mistaken about his own beliefs — he might be completely aware of his own 'non-definite belief', the one (perhaps) attributed to him by CSL. 'But I have no idea what having such a belief would be like,' you say. I agree. Neither do I, but then we are just too big ever to find out.

In other words, we are compelled to attribute to John a mistaken belief only if we are compelled to attribute to him either the belief 'the atom has spin-up in the z-direction' or the belief 'the atom has spin-down in the z-direction'. However, we are not so compelled. We may attribute to him a belief about a superposition, or even a superposed belief. Hence even if John were to exist, he need not pose a problem for CSL. Quite the contrary. John could be an invaluable tool, for he could perhaps do what thus far nobody has done — witness the collapse of a superposition.

In any case, although the debate over John rests quite heavily on extravagant speculation, it does point to an important feature of CSL, namely, its minimal attitude towards the measurement problem. The general spirit of CSL is to bring physics at the level of individual systems back into accordance with our experience. The discussion about John shows that it is *our* experience that matters, because it is our experience that seems to be violated by standard quantum theory at the individual level. For the minimalist, John's experience does not matter for the very good reason that it does not exist. If it did, then perhaps we could consult John about CSL and ask him whether it violates his experience. I do not know what he would say. However, until we are faced with the prospect of having to deal with an actual John, he cannot pose any problems for CSL.

2.4 Probabilities in orthodox interpretations

Orthodox interpretations are immediately indeterministic (by fiat), but that fact by itself does not tell us how to understand the probabilities that appear in these interpretations. Indeed, probability plays a different role in each of the two orthodox interpretations discussed here.

Because orthodoxy adopts both 'indeterminism of the moment' — 'indefiniteness', or the eigenstate–eigenvalue link — and dynamical indeterminism, an orthodox interpretation can draw on one, or both, of these forms of indeterminism to underwrite the probabilities of quantum mechanics. In a sense, standard quantum mechanics with the projection postulate focuses on the former, while CSL focuses on the latter.

Although we often think of standard quantum mechanics as altering (or better, interrupting) the deterministic evolution of Schrödinger's equation with an indeterministic evolution, perhaps 'evolution' is not the best term for what happens when we apply the projection postulate. Instead, perhaps we should conceive the matter thus: from time to time, certain events 'occur' (or, certain properties become definite) and we must then apply the projection postulate because our adherence to the eigenstate–eigenvalue link demands it. (If a system has the value a for the observable A, then its state must lie in P_a^A.) Of course, the state of a system *does* still change indeterministically, but this way of conceiving the indeterminism places less emphasis on 'indeterminisic evolution' and more on the eigenstate–eigenvalue link; the former is a *consequence* of the latter.

In CSL, on the other hand, the emphasis is properly placed on indeterministic evolution, because there we have a genuinely indeterministic *equation of motion*. Moreover, although CSL adheres to the eigenstate–eigenvalue link, its adherence is somewhat Pickwickian, because the eigenstate–eigenvalue link does not carry the consequence of 'indefiniteness' (at the macroscopic level, at least) that it does in standard quantum mechanics with the projection postulate. Indeed, given an appropriate interpretation of the wave function in CSL (e.g., the 'density of stuff' interpretation), there is no indefiniteness at any level in CSL.

Although there has been some speculation about the *origin* of probabilities in CSL (e.g., that the randomness is due to uncontrollable interaction with the 2.7 K background radiation[18]), it is clear that CSL is not *obliged* to provide any account of the origin of probabilities. Indeed, it might be undesirable to do so, because then, presumably, CSL as applied to the entire universe would be completely deterministic, so that CSL (like Bohm's theory in chapter 5) would be saddled with the problem of explaining (or explaining away) what probability means at the level of the universal wave function. Hence CSL can, and perhaps should, claim that the randomness in the CSL equation of motion is *fundamental*.

Standard quantum mechanics with the projection postulate also claims that randomness is fundamental, but much less plausibly, because here there is a *trigger* for the occurrence of randomness. You and I can *decide* whether to make a system evolve indeterministically by deciding whether to measure (or observe) it. This fact seems to demand an account. Unlike CSL, where the randomness is universal, and independent of the particularities of the world, standard quantum mechanics with the projection postulate has a form of randomness that is manifest only in special circumstances, and that fact does not sit well with the claim that collapse is a 'fundamental' process. Of course, the point I am raising (again) is just that the projection postulate lacks justification. The 'fundamental randomness' that it entails is tacked onto the theory in a most unsatisfactory way.

3

No-collapse theories

In this chapter, I consider some deterministic interpretations that nonetheless accept the eigenstate–eigenvalue link. In both cases, we will see that they solve the measurement problem through a radical interpretive move. That they must do so is clear: determinism plus the eigenstate–eigenvalue link leads immediately to the measurement problem, because it leads immediately to the attribution of states that do not assign any definite outcome (of a measurement) probability 1.

3.1 The bare theory

3.1.1 The basic idea

The bare theory is so-called because, at first glance at least, it is committed to nothing more than the eigenstate–eigenvalue link and the Schrödinger equation.[1] Of course, immediately we wonder about the measurement problem. How can the bare theory possibly square with our belief that, at the end of a measurement, the apparatus is in a definite state, *not* the state assigned to it by the bare theory?

To begin to answer this question, suppose somebody else has witnessed the measurement. You ask this person: Did you see a definite outcome? What will the person say, according to the bare theory? We may model the question as a second measurement. The quantum system begins in the state $\sum_i c_i |a_i\rangle$, and the first measurement is of the observable represented by A, whose eigenstates are $|a_i\rangle$:

$$\sum_i c_i |a_i\rangle |M_0\rangle |\Pi_0\rangle \quad \longrightarrow \quad \sum_i c_i |a_i\rangle |M_i\rangle |\Pi_0\rangle,$$

where $|\Pi_0\rangle$ is a state of the person representing 'I have not yet looked at the apparatus'. The act of witnessing the apparatus may be conceived as itself a kind of measurement:

$$\sum_i c_i |a_i\rangle |M_i\rangle |\Pi_0\rangle \quad \longrightarrow \quad \sum_i c_i |a_i\rangle |M_i\rangle |\Pi_i\rangle,$$

where $|\Pi_i\rangle$ is a state of the person representing 'I see $|M_i\rangle$'. The second measurement is your asking the person whether he or she saw a definite result. This measurement-interaction should obey

$$|a_i\rangle |M_i\rangle |\Pi_i\rangle |\text{no reply}\rangle \quad \longrightarrow \quad |a_i\rangle |M_i\rangle |\Pi_i\rangle |\text{'yes'}\rangle,$$

for any i. Then, by linearity of the Schrödinger equation, our question results in the following evolution:

$$\sum_i c_i |a_i\rangle |M_i\rangle |\Pi_0\rangle |\text{no reply}\rangle \quad \longrightarrow \quad \sum_i c_i |a_i\rangle |M_i\rangle |\Pi_i\rangle |\text{'yes'}\rangle.$$

Following the eigenstate–eigenvalue link, this final state puts the person in a definite state of saying 'yes'. If we ask: 'Did you see a definite result?', we get the answer 'yes'. Of course, most of the time the person speaks untruly, but the point of the bare theory is that the world can be as strange as we like, because what matters in the end is just whether a theory can account for the fact that we *believe* it to be definite.

The bare theory tells the very same story about statistics. Suppose that the person runs the experiment described above 100 times. Then we ask, for example: Do you have a definite belief about the percentage of times that you saw $|M_2\rangle$? According to the bare theory, the person will answer 'yes'.

3.1.2 Objections to the bare theory

However, it is far from clear that the bare theory is satisfactory, and in any case, the bare theory is not as 'bare' as it claims — it must make some substantive assumptions about the nature of consciousness and self-reflection.

Consider, for example, what happens when *you* watch the experiment yourself. In this case, you ask yourself 'Did I see a definite result?'. The bare theory must assume that this question to yourself, which is a form of introspection, is adequately modelled in the same way as the question asked to some other person, so that you will answer 'yes' to yourself, even though in fact you did *not* see a definite result.

Indeed, the bare theory must go considerably further down the road of postulated self-deception. What the bare theory cannot account for is the specific content of your belief about the result of the experiment. To see this point, suppose you ask yourself not whether you saw a definite result, but whether the result that you saw was $|M_2\rangle$. This question should probably be modelled so that you reply 'yes' if and only if the state of the system and apparatus is $|a_2\rangle|M_2\rangle$. However, then you will *never* say yes, as long as the initial state of the measured system is a superposition of eigenstates of A:

$$(c_1|a_1\rangle|M_1\rangle + c_2|a_2\rangle|M_2\rangle)|\text{no reply}\rangle$$
$$\longrightarrow \quad (c_1|a_1\rangle|M_1\rangle|\text{'no'}\rangle + c_2|a_2\rangle|M_2\rangle|\text{'yes'}\rangle).$$

In this case, you will *not* be in a definite state of saying (or thinking) anything at all. Or perhaps better, you are in an eigenstate of *some* operator, but it is not the operator corresponding to saying either 'yes' or 'no'. Just so that we have some way to refer to what you say when you are *not* in an eigenstate of the operator corresponding to saying either 'yes' or 'no', let us suppose that you say 'blah'. Hence, when you ask yourself 'Did I just see the result $|M_2\rangle$?' what you say to yourself (or, think to yourself) in reply is 'blah'.

The obvious question now is whether this account of introspection is at all plausible. Note that the process can continue indefinitely. If, after asking yourself whether you just saw $|M_2\rangle$, you ask 'Did I just say "yes"?', you will again say 'blah'. In general, you will give a definite reply *only* when you ask yourself 'Did I just give a definite reply?' Then you will say 'yes'.[2]

Now the bare theory seems to be simply wrong. Sometimes when you do the experiment, you *do* believe that the outcome was $|M_2\rangle$, or so you say. This (purported) fact of your experience the bare theory can never recover. Instead, it must claim that you are mistaken about your belief that you saw $|M_2\rangle$, and you are mistaken about your belief that you have this belief, and so on, and so on. The one belief about which you are *not* mistaken, says the bare theory, is your belief that you saw some definite result or other.

Whether we can accept the bare theory, then, depends on whether we can accept that the only belief that must be recovered by a theory is the belief *that* we saw a definite result. In general, according to the bare theory, we are radically mistaken about all of our other beliefs. In addition, of course, the world is nothing like what we think it is.

Finally, note that any adherence to the bare theory must be based on pure faith, for if the bare theory is true, then we are radically mistaken about almost everything, and in particular, we are radically mistaken about whatever (empirical) evidence we might think we have in favor of the bare theory. Hence it is not even clear that the bare theory is susceptible to rational adherence. Probably these consequences of the bare theory are sufficiently bizarre to render it unacceptable to most readers.

3.2 The many worlds and many minds interpretations

3.2.1 The central idea

The maze of many worlds and many minds interpretations of quantum mechanics is by now sufficiently serpentine to make one think twice about entering. I shall make no attempt here to sort out the details of any given approach, but instead try to articulate the central idea, and to indicate some possibilities for developing that idea. Along the way, I will mention a few authors who have pursued these developments, but my remarks here are not to be construed as a survey of existing variants on the many worlds and many minds interpretations.

Like the bare theory, the many worlds and many minds interpretations accept both the Schrödinger equation and (in a sense to be clarified) the eigenstate–eigenvalue link, and therefore, like the bare theory, they must make a radical interpretive move to steer clear of the measurement problem. Indeed, like the bare theory, these interpretations deny that measurements (in general) have a unique outcome — the one that you think you observe.

Here, however, they depart from the bare theory. Where the bare theory asserts that there is no unique outcome because there is simply no outcome, the many worlds theories and the many minds theories assert instead that *each* of the possible outcomes did in fact occur. Much of the rest of the story involves making sense of this claim, and squaring it with experience.

The person responsible for this approach to quantum mechanics is Everett, who in his 1957 paper pointed out that subsystems of a compound system in general lack their own statevectors. However, said Everett, if the compound system $\alpha \& \beta$ (composed of the two subsystems, α and β) has the state

$$|\Psi\rangle = \sum_{i,j} c_{ij}|\alpha_i\rangle|\beta_j\rangle, \tag{3.1}$$

then whenever α has the state $|\alpha_k\rangle$, β can be said to be in the 'relative state',

$$|\psi^{\beta \text{ rel } \alpha_k}\rangle = N_k \sum_j c_{kj}|\beta_j\rangle, \tag{3.2}$$

where N_k is a constant of normalization.

The probabilities for states of β that are obtained from (3.2) are exactly the probabilities generated by Lüder's rule:

$$p^{\psi^{\beta \text{ rel } \alpha_k}}(1^\alpha \otimes P^\beta) = \frac{\text{Tr}[(P_k^\alpha \otimes 1^\beta)|\Psi\rangle\langle\Psi|(P_k^\alpha \otimes 1^\beta)P^\beta]}{\text{Tr}[|\Psi\rangle\langle\Psi|(P_k^\alpha \otimes 1^\beta)]}, \tag{3.3}$$

where $P_k^\alpha = |\alpha_k\rangle\langle\alpha_k|$, 1^α is the identity on $\mathcal{H}^\alpha$ (the Hilbert space for α), and similarly for 1^β.

Having made these (or similar) observations, Everett then writes:

There does not, in general, exist anything like a single state for one subsystem of a composite system. Subsystems do not possess states that are independent of the states of the remainder of the system, so that the system states are generally *correlated* with one another. One can arbitrarily choose a state for one subsystem, and be led to the relative state for the remainder.[3]

The meaning of Everett's words is perhaps not entirely clear — and further clarification is not forthcoming in the remainder of the paper — but apparently we are to imagine that, for example, α has no particular state, but instead 'has' *each* of the $|\alpha_i\rangle$, and that for each i, β 'has' $|\psi^{\beta \text{ rel } \alpha_i}\rangle$ *relative to* α's having $|\alpha_i\rangle$. The most simple-minded way to make sense of this idea is to imagine that α exists in many worlds, in each of which α has just one of the $|\alpha_i\rangle$ and β has the corresponding relative state.

Now, it may be that we can make sense of Everett's words *without* having to posit a multiplicity of worlds. Saunders, for example, seems to take this view.[4] Nonetheless, I shall stick with the idea that there is a multiplicity of worlds, so that the sense attached to a statement like 'α does not possess one of the $|\alpha_i\rangle$' is: α possesses *each* of the $|\alpha_i\rangle$, one in each world.

There are two assymetries in the account as given thus far. First, there is an assymetry between α and β — in each world, α has one of the $|\alpha_i\rangle$ while β has a state only *relative* to α. Second, the entire construction has been performed only in the context of a given basis, $\{|\alpha_i\rangle, |\beta_j\rangle\}$, for $\mathcal{H}^\alpha \otimes \mathcal{H}^\beta$.

In each case, there is an obvious way to regain symmetry (and thereby avoid having to explain the assymetry). In the first case, simply add more worlds, one for each $|\beta_j\rangle$. In *these* worlds, β has one of the $|\beta_j\rangle$ while α has the corresponding *relative* state, $\sum_i c_{ij}\alpha_i$, suitably normalized. In the second case, we may again add more worlds, indeed, one *set* of worlds for *each* basis of $\mathcal{H}^\alpha \otimes \mathcal{H}^\beta$.

Both of these modifications lead to difficulties, as I will discuss in the next section. Hence one might prefer, instead, to try to justify the assymetries. In Everett's paper, the assymetry between α and β is apparently meant to be justified on the grounds that α is an 'observer' while β is the 'observed' system. (The claim that the relation between α and β is one of 'observation' is, of course, compatible with the proposed restoration of symmetry between α and β — in that case one would say that each 'observes' the other.) However, apart from the difficulty of saying what constitutes an 'observer', this view apparently renders the many worlds interpretation itself unnecessary. If we have a satisfactory account of which systems count as observers, and what their 'pointer' states are (which involves explaining the second assymtery too), then the projection postulate is well defined, and the trappings of the many worlds interpretation is superfluous.

Given this justification of the assymetry between α and β, the assymetry among bases can be justified in a similar way: the 'correct' basis in which to expand $|\Psi\rangle$ ($\in \mathcal{H}^\alpha \otimes \mathcal{H}^\beta$) is the one that includes the apparatus' pointer states and the eigenstates of the measured observable.

Given some 'preferred basis', there is another way to avoid the first assymetry between α and β, which is to forget about relative states. Instead, expanding $|\Psi\rangle$ in the preferred basis, as in (3.1), we may say that for each term, $c_{ij}|\alpha_i\rangle|\beta_j\rangle$, there is one world (or, class of worlds), in which α has $|\alpha_i\rangle$ and β has $|\beta_j\rangle$. Of course, this view still faces the problem of defining the preferred basis. In addition, once such a basis is found, this view, like the others, faces the question of why we cannot then simply adopt the projection postulate.

3.2.2 Many minds

Recall the 'minimalism' of CSL: it sought to recover our *experience* of the world, and was willing to allow indefiniteness in the world, so long as our experience of it remains definite. There is a variant of the many worlds interpretation that adopts a similar attitude. This variant is the many minds interpretation.[5]

The many minds interpretation may be characterized by how it solves the preferred basis problem. (The many minds interpretation resolves the other assymetry by ignoring relative states in the way described at the end of the previous section.) The assumption is that there is *some* basis that describes definite states of belief for everybody (or at least, every mind) in the universe. Call the elements in this basis $|\eta_i\rangle$. Then, if $|\Psi\rangle$ is the statevector of the universe, we write $|\Psi\rangle$ in the $|\eta_i\rangle$-basis: $|\Psi\rangle = \sum_i c_i|\eta_i\rangle$. Then we say that each person (or, mental agent) in fact has many minds (or perhaps, classes of minds), one for each element in $\{|\eta_i\rangle\}$.

In other words, each $|\eta_i\rangle$ describes a definite state of belief for *every* mental agent in the universe. (The theory makes no commitment about how many such agents there are.) However, because the universe 'has' all of the $|\eta_i\rangle$, each agent 'has' many beliefs, or perhaps better, to each agent there corresponds many minds.

It will be helpful for later to see why agents will always believe themselves to agree about the results of a commonly witnessed measurement. Suppose that you and I both witness an experiment (recall the discussion of the bare theory for notation):

$$\sum_i c_i |\alpha_i\rangle |M_0\rangle |\Pi_0^{\text{you}}\rangle |\Pi_0^{\text{me}}\rangle$$
$$\longrightarrow \sum_i c_i |\alpha_i\rangle |M_i\rangle |\Pi_i^{\text{you}}\rangle |\Pi_i^{\text{me}}\rangle.$$

Now, you as believing agent have many minds, one for each $|\Pi_i^{\text{you}}\rangle$, and similarly for me. However, clearly there is some sense of 'you' (and 'me') according to which 'you' are aware of just *one* of these minds — call it 'you as mind'. (There are many 'yous' in this sense.) Equally clearly, you as mind and me as mind need *not* have the same belief about the result of the measurement. So why is it that when 'we' (as minds) meet, 'we' (as minds) believe that we *do* agree?

The answer is in the equations. Suppose 'I' ask 'you' whether 'you' believe what 'I' do about the state of the measuring apparatus. That interaction (i.e., the interaction between 'you' and 'me' as 'I' ask 'you' this question) should obey

$$|\Pi_i^{\text{you}}\rangle |\Pi_j^{\text{me}}\rangle |\text{no reply}\rangle \longrightarrow \begin{cases} |\Pi_i^{\text{you}}\rangle |\Pi_j^{\text{me}}\rangle |\text{'yes'}\rangle & \text{if } i = j \\ |\Pi_i^{\text{you}}\rangle |\Pi_j^{\text{me}}\rangle |\text{'no'}\rangle & \text{if } i \neq j \end{cases}.$$

Then, adding *this* interaction onto the end of the measurement, we have:

$$\sum_i c_i |\alpha_i\rangle |M_i\rangle |\Pi_i^{\text{you}}\rangle |\Pi_i^{\text{me}}\rangle |\text{no reply}\rangle$$
$$\longrightarrow \sum_i c_i |\alpha_i\rangle |M_i\rangle |\Pi_i^{\text{you}}\rangle |\Pi_i^{\text{me}}\rangle |\text{'yes'}\rangle.$$

so that in fact 'we' always *believe* that 'we' agree. The many minds interpretation therefore shares with the bare theory the possibility that in general you and I are radically mistaken about the beliefs of others, though it differs from the bare theory in that 'you' are always correct about 'your' own beliefs.

Note that the many worlds interpretation does not tell the same story about agreement. The notion of 'you (as mind)' is not part of the many worlds interpretation. Instead, there are 'you as part of a given world' and 'me as part of a given world'. However, while 'you as mind' and 'me as mind' can interact even if our beliefs (about the measuring device) differ — because there is just one world, and it contains *all* minds — you as part of a given world and me as part of a given world cannot interact unless the worlds in question are the same world. However, then we *will* agree about the state of the measuring device, because in a given world, it has just one state.

To put the point more generally, in the many minds interpretation there is just one world, but many minds. If two disagreeing minds should interact, they will *take* themselves to agree — they will agree about agreement. In the many worlds interpretation, there are many worlds, but each believing agent has just one mind in any given world; and believing agents automatically agree in any given world.

3.3 The consistent histories approach

The consistent histories approach to quantum mechanics — due originally to Griffiths[6] — is perhaps the least developed interpretation of those considered in this book, if indeed it may be called an 'interpretation'. Hence I am not even sure whether it belongs in this chapter. It has, however, commanded considerable attention in the past few years, and seems worth some mention. In this section I shall say just a little about the consistent histories approach, although, as will become clear, it is not obvious what the approach claims about the world.

3.3.1 The formalism of consistent histories

To discuss the consistent histories approach, it is easiest to work in the Heisenberg picture, so that observables are time-dependent. In particular, recall, the projections corresponding to various properties are time-dependent and their evolution is given by:

$$P(t) = U_{0,t}P(0)U_{0,t}^{-1}, \tag{3.4}$$

where $P(0)$ is the projection at an arbitrarily chosen origin and $U(t)$ is the system's evolution operator from time 0 to time t.

The central concept in the consistent histories approach is the *history*, a time-ordered sequence of properties (increasing index always indicates increasing time):

$$P(t_1) \to P(t_2) \to \cdots \to P(t_f). \tag{3.5}$$

Probabilities in this approach are assigned to histories according to the rule that the probability of the history $P(t_1) \to P(t_2) \to \cdots \to P(t_n) \to P(t_f)$ is

$$\text{Tr}[P(t_f) \cdots P(t_2)P(t_1)W_0 P(t_1)P(t_2) \cdots P(t_n)P(t_f)]. \tag{3.6}$$

(I assume henceforth that $W(0)$ and $P(t_f)$ represent physically possible initial and final states of a single quantum system, i.e., $\text{Tr}[W(0)P(t_f)] \neq 0$.) As do many other interpretations, the consistent histories approach aims to obey the classical rules of probability. To formulate the relevant requirement, we need the notion of a family of histories.

A history can be embedded in various *families*, where a family is denoted

$$W(0) \to \{P^{(a_1)}(t_1)\} \to \{P^{(a_2)}(t_2)\} \to \cdots \to \{P^{(a_n)}(t_n)\} \to P(t_f). \tag{3.7}$$

Each $\{P^{(a_k)}(t_k)\}$ is a set of projections that constitutes a resolution of the identity operator on $\mathscr{H}$. That is, for each k,

$$\sum_{a_k} P^{(a_k)}(t_k) = 1. \tag{3.8}$$

The classicality requirement is imposed on families of histories, any family meeting the requirement being called a 'consistent family'.[7] A *consistent family of histories* is one for which, for every k, $1 \le k \le n$, every $P^{(l)}(t_k), P^{(m)}(t_k) \in \{P^{(a_k)}(t_k)\}$, and every history in the family:

$$\begin{aligned}
\mathrm{Tr}\Big[& P(t_f)P^{(a_n)}(t_n) \cdots \big(P^{(l)}(t_k) + P^{(m)}(t_k) \big) \cdots P^{(a_1)}(t_1) W(0) P^{(a_1)}(t_1) \\
& \cdots \big(P^{(l)}(t_k) + P^{(m)}(t_k) \big) \cdots P^{(a_n)}(t_n) P(t_f) \Big] \\
= \mathrm{Tr}\Big[& P(t_f)P^{(a_n)}(t_n) \cdots P^{(l)}(t_k) \cdots P^{(a_1)}(t_1) W(0) P^{(a_1)}(t_1) \\
& \cdots P^{(l)}(t_k) \cdots P^{(a_n)}(t_n) P(t_f) \Big] \\
+ \mathrm{Tr}\Big[& P(t_f)P^{(a_n)}(t_n) \cdots P^{(m)}(t_k) \cdots P^{(a_1)}(t_1) W(0) P^{(a_1)}(t_1) \\
& \cdots P^{(m)}(t_k) \cdots P^{(a_n)}(t_n) P(t_f) \Big]. \tag{3.9}
\end{aligned}$$

That is, a consistent history is one for which the usual sum rule of classical probability holds. For example, if there are only two mutually exclusive ways to get from the ballroom to the conservatory — via the kitchen or via the study — then classically we would expect that the probability of getting to the conservatory given that one starts at the ballroom is equal to the probability of getting to the conservatory given that one starts at the ballroom and goes through the kitchen, plus the probability of getting to the conservatory given one starts at the ballroom and goes through the study. (This condition fails, for example, in the two-slit experiment.) Consistent families of histories are therefore histories in which interference effects can be neglected. Finally, a history is called a *consistent history* just in case the smallest family of histories in which it can be embedded is a consistent family. Two histories are called *incompatible* just in case they cannot be simultaneously embedded in a consistent family of histories.

3.3.2 Interpretation of the formalism

The formalism of consistent histories is a modification of quantum probability. Instead of $L_{\mathscr{H}}$ as the algebra of events, the formalism of consistent histories begins with the space of all histories. Over this space, the formalism uses the usual quantum-probabilistic measure, p^{W_0}, as can be seen from (3.6). The resulting probability theory is non-classical, in the same ways that quantum probability is non-classical. However, if we restrict attention to some consistent families of histories — i.e., if we take as the sample space not all histories, but all histories from some consistent family — then the rules of classical probability are recovered.

However, mere definition of a classical probability space is not enough to give the formalism an unambiguous physical meaning. We need, of course, to make the usual connection between projection operators and physical quantities. But we need more. Here is the test that we want the formalism to meet: Given a state, W_0, does the formalism yield a probability for the occurrence of an event, $P(t)$? Thus far, the answer is 'no', even if $P(t)$ is known to be possibly-occurrent (and in this respect, the present approach differs from modal interpretations, discussed in chapter 4).

The reason is that the formalism does not provide probabilities for events, but for histories. Now, the apparently easy solution is to allow that the event $P(t)$ can be considered a 'degenerate' history, a history with only one event. In that case, the formalism will give exactly the right probability for $P(t)$. But what has happened to our probability space? The algebra of histories is just $\{0, P(t), P^{\perp}(t), 1\}$, and the formalism is now crippled — it can make predictions *only* about the occurrence of $0, P(t), P^{\perp}(t)$, and 1. Perhaps then we should say that the probability of $P(t)$ is the probability that it has given *any* history containing $P(t)$. The problem here is that there is no unique such probability — the probability of $P(t)$ will differ, depending on which consistent family we use as our algebra of histories.

For Griffiths, the way around these problems (though he does not raise them in the way I do here) is to adopt a 'perspectivalism' about the formalism. Although we ask for the probability that $P(t)$ occurs, given W_0, our question is, in Griffiths' view, ambiguous. Instead we should ask: *relative to a given consistent family of histories, what is the probability that $P(t)$ occurs, given W_0?* This question is answerable by the consistent histories formalism. However, what gives rise to this perspectivalism? Why *must* I specify a 'perspective' (a consistent family) in order that my questions about $P(t)$ be answerable by the theory? I raise these questions again below.

3.3.3 Is the consistent histories approach satisfactory?

3.3.3.1 Consistent histories and macroscopic experience

Let us assume that Griffiths' perspectivalism is satisfactory for the moment, and that there is a well-defined way to decide what one's perspective is, or ought to be (i.e., a well-defined way to choose a consistent family relative to which questions about events are asked). Will the approach be satisfactory? Will it describe our macroscopic and classical experience?

The (unsatisfying) answer seems to be: Only if we take that experience as given already. The reason is that without knowing ahead of time what the actual history is, we will not, in general, be able to choose a consistent family that contains that history. A simple example will suffice to make this point. Suppose we are given the actual history of the world up to time t_1, and somehow we find the (or a) 'correct' perspective from which to describe this history, i.e., a consistent family of histories one of whose elements is the actual history. Now, suppose that we do not know how the world will evolve from time t_1 to time t_2. How will we choose a consistent family of histories from time 0 to time t_2 that is *guaranteed* to include the actual history? It is difficult to see how we can do so, without knowing how the world will evolve — there are too many 'wrong' choices.

Consider, for example, the following realistic situation. A spin-1/2 particle approaches a Stern–Gerlach device, which will measure either σ_z or σ_x. Whether it measures one or the other is not decided until time t', where $t_1 < t' < t_2$. In this situation, it is not possible to specify at t_1 a consistent family of histories up to t_2 that is guaranteed to include the measurement-result. The reason is that in order to guarantee that it includes the result, this family would have to contain histories with $P_+^{\sigma_z}(t_2), P_-^{\sigma_z}(t_2), P_+^{\sigma_x}(t_2)$, and $P_-^{\sigma_x}(t_2)$. (Here $P_+^{\sigma_z}(t_2)$ is just the time-dependent version of $P_+^{\sigma_z}$, where σ_z represents spin in the z-direction, as usual.) In that case, however, the resulting family will not be consistent, as is easily checked.

Barring a solution to this problem, the formalism of consistent histories provides a way of telling a consistent story about only those events that we already know to have occurred. However, there are at least two approaches to solving the problem. One is to begin the very ambitious program of finding a criterion for the selection of families of histories that will always include the actual history. A detailed theory of human perception, for example, might provide a criterion that is at least good enough to select families containing the history of our perceptions. However, no such criterion is available now, and does not seem to be just around the corner. (Note the similarity of this problem to the problem of finding a preferred basis for the many worlds, or indeed, the many minds, interpretation.)

The second approach is to allow that in some sense *every* history is legitimate. This approach seems to be Griffiths' own way of looking at the matter. It is similar to the version of the many worlds interpretation in which every basis represents a set of worlds, and faces the same problem (concerning probabilities) that I discuss below.

3.3.3.2 Inconsistency of different consistent histories

Griffiths' approach is perspectival not only in the sense that questions about probabilities must be relativized to a consistent family, but also in the (more radical) sense that probabilities from different perspectives may not be compared.

To see why, consider a standard EPR–Bohm experiment, in which two particles (α and β) leave a source in the singlet state:

$$|\Psi_{\text{singlet}}\rangle = \frac{1}{\sqrt{2}}\left(|z,+\rangle^\alpha|z,+\rangle^\beta - |z,-\rangle^\alpha|z,+\rangle^\beta\right), \tag{3.10}$$

where $|z,+\rangle^\alpha$ indicates spin-up in the z-direction for α, and so on. Each particle has its spin measured by a Stern–Gerlach apparatus set to measure spin in the i and j directions, respectively. Let the projection representing the event 'the detector for α indicates spin-up [spin-down] in the z-direction' be denoted '$D^\alpha_{z,+}$' ['$D^\alpha_{z,-}$'], and similarly for '$D^\beta_{z,+}$' ['$D^\beta_{z,-}$']. Finally, let the spin operators σ carry a superscript to indicate whether they are on the α or β system. (In what follows, I assume that the measurements are perfectly accurate.)

Now, consider the family of histories ($i = z$, $j = z'$)

$$
\begin{aligned}
|\Psi_{\text{singlet}}\rangle\langle\Psi_{\text{singlet}}| \otimes \mathbf{1}^D \;\rightarrow\; &\Big\{ P^{\sigma^\alpha_z}_+(t_1) \otimes P^{\sigma^\beta_{z'}}_+(t_1) \otimes \mathbf{1}^D, \\
& P^{\sigma^\alpha_z}_+(t_1) \otimes P^{\sigma^\beta_{z'}}_-(t_1) \otimes \mathbf{1}^D, \\
& P^{\sigma^\beta_z}_-(t_1) \otimes P^{\sigma^\beta_{z'}}_+(t_1) \otimes \mathbf{1}^D, \\
& P^{\sigma^\beta_z}_-(t_1) \otimes P^{\sigma^\beta_{z'}}_-(t_1) \otimes \mathbf{1}^D \Big\} \\
\rightarrow\; &\Big\{ \mathbf{1}^{\alpha\&\beta} \otimes D^\alpha_{z,+}(t_2) \otimes D^\beta_{z',+}(t_2), \\
& \mathbf{1}^{\alpha\&\beta} \otimes D^\alpha_{z,+}(t_2) \otimes D^\beta_{z',-}(t_2), \\
& \mathbf{1}^{\alpha\&\beta} \otimes D^\alpha_{z,-}(t_2) \otimes D^\beta_{z',+}(t_2), \\
& \mathbf{1}^{\alpha\&\beta} \otimes D^\alpha_{z,-}(t_2) \otimes D^\beta_{z',-}(t_2) \Big\} \\
\rightarrow\; & \mathbf{1}^{\alpha\&\beta} \otimes \mathbf{1}^D,
\end{aligned}
$$

where $\mathbf{1}^{\alpha\&\beta}$ is the identity operator on the Hilbert space for the compound system composed of α and β and $\mathbf{1}^D$ is the identity operator for the compound system composed of the pair of detectors, one for each system. (Putting the identity at the end of the history is only a calculational convenience — because it is certain to occur, it adds no content to the history.) One can show that this family is consistent.[8] A straightforward calculation yields that, relative to this family of histories, the probability of $P^{\sigma^\alpha_z}_+(t_1)$ given $D^\alpha_{z,+}(t_2) \otimes D^\beta_{z',+}(t_2)$ is 1.

On the one hand, this result is exactly what one wants in a 'realistic' account such as Griffiths hopes to provide.[9] It says that the value indicated by the α-detector after the spin-measurement at t_2 was in fact possessed by α prior to the measurement. On the other hand, this result leads to the following oddity. Consider a second (consistent) family of histories, identical to the first except that the events at t_1 are

$$\{P_+^{\sigma_{z'}^\alpha}(t_1) \otimes P_+^{\sigma_{z'}^\beta}(t_1) \otimes \mathbf{1}^D,$$
$$P_+^{\sigma_{z'}^\alpha}(t_1) \otimes P_-^{\sigma_{z'}^\beta}(t_1) \otimes \mathbf{1}^D,$$
$$P_-^{\sigma_{z'}^\alpha}(t_1) \otimes P_+^{\sigma_{z'}^\beta}(t_1) \otimes \mathbf{1}^D,$$
$$P_-^{\sigma_{z'}^\alpha}(t_1) \otimes P_-^{\sigma_{z'}^\beta}(t_1) \otimes \mathbf{1}^D\}. \tag{3.11}$$

One can then show that the the probability of $P_+^{\sigma_{z'}^\alpha}(t_1)$ given $D_{z,+}^\alpha(t_2) \otimes D_{z',+}^\beta(t_2)$ is 1.

It appears that we have derived that, given $D_{z,+}^\alpha(t_2) \otimes D_{z',+}^\beta(t_2)$ (which we may suppose to have actually occurred), $P_+^{\sigma_z^\alpha}(t_1)$ *and* $P_+^{\sigma_{z'}^\alpha}(t_1)$ occur with certainty. Griffiths' reply is the following (where, in his notation, 'p_1' is '$P_+^{\sigma_z^\alpha}(t_1)$' and '$u_1$' is '$P_+^{\sigma_{z'}^\alpha}(t_1)$'):

The inference from the correctness of "p_1" and "u_1" individually to the correctness of the (meaningless) "p_1 and u_1" is blocked in the consistent history approach by the rule ... that probability calculations and logical inferences ... *must* be carried out in the context of a *single* family of consistent histories.[10]

Griffiths argues that this radical perspectivalism follows from a view of inference in which 'p_1' and 'u_1' does not entail 'p_1 and u_1'. Of course, those who are uncomfortable with perspectivalism will hardly be relieved to hear that it follows from a facile revision of logic, and we may well ask: *Why may we not compare probabilities from different families of histories?*

I cannot see any obvious answer to this question from within the formalism of consistent histories. At the very least, it seems to put every family of histories on a par — no family is 'the correct' family from which to compute probabilities. Instead, they are all equally 'legitimate' frameworks from which to calculate probabilities. Moreover, within a given family, there is no mechanism for selecting just one as occurrent — i.e., proponents of this approach do not seem to endorse the projection postulate, and do not suppose that just *one* history is selected as 'actual'.

These points should make it clear why I have included a discussion of this approach in this chapter. Although it is in a sense indeterministic — *probabilities* are assigned to different histories — the approach apparently selects no particular history as 'occurrent'. Similarly, while the many worlds interpretation assigns probabilities to worlds (though I will not discuss these probabilities until the next section), it does not select any particular world as 'occurrent' (they are all occurrent), so that the totality of worlds, like the totality of possible histories, is simply given by the (deterministically evolving) quantum-mechanical statevector for the universe.

Seen from this point of view, the approach in terms of consistent histories appears to differ little from the many worlds interpretation. Indeed, the difference seems to be only that while the many worlds interpretation allows, in principle, worlds whose history is (in the technical sense) inconsistent, the approach in terms of consistent histories obviously allows only consistent histories. Apart from that minor point, they seem to face the same questions and problems (plus the challenge of explaining why histories should be consistent[11]).

3.4 What is 'interpretive minimalism' and is it a virtue?

Advocates of the bare theory, consistent histories, and, most especially, the many worlds interpretation are often heard to claim in their favor that these theories 'add nothing' (or, 'as little as possible') to quantum mechanics. Or sometimes they will say that these interpretations 'add nothing to the physics'.

Such claims are clearly meant to garner allegiance to these interpretations, but little is said about *why* they should do so. In particular, it is not even clear what is being claimed for these interpretations, and why it is a virtue anyway.

There are two obvious strategies for making the argument that the many worlds interpretation adds little or nothing to quantum theory as it stands. First, one might argue for little or no distinction between physics and metaphysics, or formalism and interpretation, hence hinting that these interpretations can simply be 'read off of' the formalism of quantum probability theory, so that any other interpretation of the same formalism must clearly be adding something not already there. Second, one might argue for a significant distinction between physics and metaphysics, or formalism and interpretation, hence suggesting that these interpretations 'add nothing' to the formalism in the sense of their 'adding no new physics'.

Neither strategy shows much hope of being convincing. Beginning with the first, we may note immediately that the many worlds and many minds interpretations accept the eigenstate–eigenvalue link, in the sense that a given system cannot be said absolutely (i.e., in every world) to possess a value, *a*, for the observable, A, unless it is in a state in P_a^A. However, on what basis can we say that the eigenstate–eigenvalue link is already a part of quantum probability theory? Is it contained in the very notion of 'probability' that only properties with probability 1 are possessed? On some interpretations of probability, perhaps it is, but clearly there are some interpretations of probability that do not carry this implication. Hence the acceptance of the eigenstate–eigenvalue link already seems to rely on certain interpretations of probability theory, and it is pretty clear that probability theory is not susceptible to just one interpretation.

Advocates of this strategy are likely to retort that I have already skewed the issue by framing it in terms of whether the many worlds and many minds interpretations can be read off of quantum *probability* theory, rather than quantum *mechanics*. They might suggest that the eigenstate–eigenvalue link is plausibly denied only when we have already agreed to interpret the quantum state as a probability measure. To adopt *that* interpretation, however, is already to go beyond what quantum mechanics says. Quantum mechanics provides a state for every system, and says how that state evolves. It is *that* theory, they say, that leads inevitably to the many worlds interpretation.

However, if any interpretation has the right to call itself 'as close as possible' to quantum mechanics without the projection postulate, it would seem to be the bare theory. However, even the bare theory had to add to quantum mechanics some substantive assumptions (about the nature of belief, introspection, and our experiential access to the world). In any case, the very existence of a plausible competitor for the title 'as close as possible to the physics' strongly suggests that the many worlds and many minds interpretations are not simply contained in quantum mechanics.

In any case, the eigenstate–eigenvalue link is clearly not the only element added to quantum mechanics by the many worlds and many minds theories. For example, the notion of a relative state, although expressible in terms of the formalism of quantum mechanics, is no part of quantum mechanics. (After all, physicists did quantum mechanics for thirty years before the notion of a relative state was introduced.) For that matter, where in quantum mechanics do we find the notion of a 'world', or a 'mind'?

Consider now the second strategy: arguing that there is some distinction between physics and metaphysics, and that the many worlds and many minds interpretations add nothing to the physics. The most obvious place to *try* to draw the distinction between physics and metaphysics is at the level of experiment. That is, any 'interpretation' of quantum mechanics that is experimentally distinguishable from quantum mechanics should, we may suppose, be said to 'add new physics'.

There are two problems with this proposal. First, it is not clear that quantum mechanics itself provides an unambiguous answer to every experimental question. Second, there are plenty of interpretations that could claim to add no new physics, by this definition.

Let us focus on the second problem. Suppose that 'adding no new physics' is a virtue, and suppose that 'adding no new physics' means 'making no experimental predictions beyond those of quantum mechanics' (and suppose that the latter is well defined!). Still, it is not at all clear that the interpretations discussed in this chapter have a monopoly on this virtue. Indeed, most interpreters of quantum mechanics take it as a *desideratum* to be experimentally equivalent to quantum mechanics, insofar as the latter is unambiguous. It is at least unclear that the many worlds and many minds interpretations succeed and the others fail in this regard.

However, perhaps those who claim 'no new physics' as a virtue for these interpretations have something else in mind. If so, then it is not clear why 'no new physics' is a virtue at all. After all, we trust quantum mechanics as it happens to be formulated primarily because it is empirically very successful, and provides some measure of explanation for certain physical phenomena. Suppose, however, that some *other* theory were equally successful, and were equally explanatory. To reject it because it is not the same as quantum mechanics is, it seems, to be too much attached to the particular historical circumstances that gave rise to the formulation of quantum mechanics. There is nothing magical about the Schrödinger equation. It is an equation that a person — Erwin Schrödinger — wrote down, and we trust it because it is successful. If some *other* equation is equally successful, then the only reason we have left for preferring the Schrödinger equation is that a man named 'Erwin Schrödinger' wrote it down.

Of course, this entire discussion presupposes that 'quantum mechanics' is well defined in the first place, and part of the presupposition of this book is that it is not. There is, of course, 'quantum probability theory plus the projection postulate', which passes for 'standard quantum mechanics' in most textbooks, but no advocate of the interpretations discussed in this chapter thinks that we must maintain adherence to *that* interpretation. There is quantum probability theory alone, but it is merely a mathematical theory, and by itself has nothing to do with experiments or 'physics'. From this point of view — the point of view adopted here — the claim that a given interpretation 'adds nothing' to the physics of quantum mechanics does not even make sense. At the best, it could mean only that a given interpretation does not require a change in what is presupposed by the practice of working quantum physicists. However, apart from the extreme difficulty of saying just what that practice is, it is not clear why adding nothing to it is a virtue. (Nor is it clear, again, that other interpretations would fail this test.)

So my own conclusion is that this argument in favor of the many worlds and many minds theories is not convincing. When 'adding no new physics' is rendered in such a way as to make it clearly a virtue, it becomes a virtue shared by many interpretations. When it is rendered in such a way that only a few interpretations do it, it is no longer clearly a virtue. And worse, because 'quantum mechanics' is already ill defined, it is not clear to what we are 'not adding' in the first place.

Of course, it might turn out that only a radical move such as those suggested by these interpretations can escape the problems facing any attempt to turn quantum probability theory into a physical theory. However, *that* sort of argument in their favor can only be made by means of a careful evaluation of all alternatives. While this book in no way pretends to be such an evaluation, we will see in subsequent chapters some important alternatives to the many worlds and many minds theories, alternatives that cannot be easily dismissed.

3.5 Probabilities in no-collapse interpretations

The consistent histories interpretation is evidently indeterministic in general. It is possible to find families of histories only one of whose elements has non-zero probability, but of course in general, and for interesting families of histories, more than one history will have non-zero probability. Indeed, I have already mentioned that this fact seems to lead the consistent histories interpretation in the direction of the many worlds (or many minds) interpretation.

However, *defining* a probability measure and making sense of it are two different things. In addition, although the consistent histories formalism apparently provides some way to define probabilities for worlds, in fact it papers over some serious problems. For the reasons already mentioned above, it seems that the best way to make sense of the formalism of consistent histories is in terms of many worlds (or many minds), and if so, then the approach in terms of consistent histories faces the same problems making sense of probability that are already faced by the many worlds and many minds interpretations.

Indeed, in the discussion of many worlds and many minds interpretations, I did not mention probability, and that omission is no accident — probability does not enter into these interpretations in any natural way, or at least not in any obvious way. The problem is just that the interpretations are completely deterministic. So where is there room for probability? Why does the quantum world *seem* to us to be probabilistic?

One sort of answer is immediately evident.[12] First, we may try to define some notion of a world over time. Note that the many worlds and many minds interpretations as I described them included no notion of identity across time of a world (or mind). If, however, we could find some way to incorporate such a notion into these interpretations, then we could, perhaps, find (or at least define) a *dynamics* for worlds (or minds), plus an initial distribution over worlds (or minds) that reproduces all of the single-time probabilities of standard quantum mechanics. Indeed, at least one such proposal is obvious (assuming that a preferred basis has been found): let the initial distribution over worlds (or minds) be the quantum-mechanical distribution, and let the transition probabilities be given be

$$p\Big(|\eta'_j\rangle \text{ at } t' \big| |\eta_i\rangle \text{ at } t\Big) = p\Big(|\eta'_j\rangle \text{ at } t'\Big), \tag{3.12}$$

where $t' > t$ and $|\eta_i\rangle$ and $|\eta'_j\rangle$ are elements of the preferred basis at times t and t', respectively. In this case, in every world, the probability distribution over elements of the preferred basis is exactly the quantum-mechanical distribution. Of course, such a dynamics is quite unintuitive. On the other hand, the formalism of consistent histories might provide a more satisfactory dynamics. Whether it can do so depends on whether the 'histories' provided by the many worlds or many minds interpretations are consistent. (Another possibility would be to adopt the dynamics for modal interpretations, discussed in chapter 4.)

Although the claim has sometimes been made (e.g., by Everett himself) that the many worlds interpretation alone *entails* the usual probability calculus of quantum mechanics, I do not see how it does. It seems that *some* account such as that of the previous paragraph is required. In particular, without a notion of identity across time of a world (or mind), it is unclear how probabilties can be made empirically manifest; i.e, the connection between probabilities and relative frequencies (over time) is severed. Indeed, the very notion of performing an experiment (which inevitably takes time) is apparently unavailable without the prior notion of what constitutes *the same world* (or mind) over time.

Of course, a notion of identity across time is not sufficient to determined the probabilities of quantum mechanics, or indeed any probabilities. I say 'of course', but contrary arguments of two sorts have been made. First, it is sometimes suggested that the many worlds (or minds) interpretation is committed to a *flat* distribution over worlds. Why? Because if what exists, fundamentally, is a set of worlds, then the *a priori* distribution over them should assign each world equal probability.

If there are only finitely many worlds, N, then it does seem quite natural to suppose that the *a priori* probability of each world is $1/N$. (That is, the *a priori* chance that 'you' are in any given world is $1/N$.) However, if there are continuously many worlds, then each of them *automatically* has probability zero,[13] and hence equal probability, but this fact is compatible with *many* probability densities over the set of worlds. Hence if there are continuously many worlds, the many worlds (or minds) interpretation is *not* committed to a flat probability distribution over worlds.[14]

On the other side, some authors — including, apparently, Everett — have supposed that the quantum-mechanical probabilities follow immediately from the principles of the many worlds interpretation. However, many authors have noted that the purported derivation of the quantum-mechanical probabilities is circular,[15] and the argument of the previous paragraph shows that they must be right.

So there are *two* 'problems of probability' in the many worlds (or minds) interpretation. The first is how to make *sense* of probability in the first place. The second is how to justify the *quantum-mechanical* probabilities.

The first problem appears to admit two types of solution. The type that I mentioned is that of describing an indeterministic evolution of each world over time. A second type of solution would regard all probabilities as epistemic, as probability measures over which world 'I' am in (or, over which mind is 'mine'). In both approaches, the underlying difficulty is that of defining a notion of identity, a distinctly philosophical problem whose solution, if there is one, has yet to be found.

As far as I can tell, the second problem of probability has no interesting solution. However, this problem is not faced *only* by many worlds interpretations. Indeed, it seems we can *always* ask 'why are probabilities given by *that* measure?', and it is not clear to me how there could ever be a good answer to this question.

4

Modal interpretations

4.1 The quantum logic interpretation

4.1.1 The basic idea

I have already hinted at the quantum logic interpretation in chapter 1. The basic idea is to take the lattice-theoretic operations of meet ($\wedge$), join ($\vee$), and orthocomplement ($^\perp$) as the 'true' logical operations: meet is 'and', join is 'or', and orthocomplement is 'not'. In other words, the logic that the world obeys is quantum logic. To assess the truth of the statement 'P or Q', we must represent it as '$P \vee Q$', and so on.

What consequence does this move have for the eigenstate–eigenvalue link? Well, let us assume (as proponents of quantum logic would have us do) that the statement 'A has some definite value' is equivalent to 'A has the value a_1, or A has the value a_2, ...'. (There is one disjunct for each eigenvalue of A.) The quantum-logical representation of this statement is

$$P_{a_1}^A \vee P_{a_2}^A \vee \cdots \tag{4.1}$$

or

$$\bigvee_i P_{a_i}^A . \tag{4.2}$$

However, the eigenspaces of any operator span the entire Hilbert space. Therefore, $\bigvee_i P_{a_i}^A \equiv \mathbf{1}$. And $\mathbf{1}$ is a tautology — it is the always-true proposition in quantum logic. (This much of the quantum logic interpretation is true in the orthodox interpretation too: *every* state assigns $\mathbf{1}$ probability 1.) Therefore, the statement 'A has some definite value' is a tautology in quantum logic. Of course, the argument does not depend on which A we choose: *every* observable has a definite value, according to the quantum logic interpretation. Therefore the eigenstate–eigenvalue link must fail in the quantum logic interpretation, because no quantum state can assign probability 1 to some value for each observable.

Proponents of the quantum logic interpretation will sometimes speak otherwise.[1] Rather than denying the eigenstate–eigenvalue link, they will suppose instead that a system has *many* 'quantum states', one for each observable. However, the difference between their way of putting the point and mine makes no real difference. What they will admit is that we can assign at most a single quantum-mechanical state at a time to a system (even if it actually has many quantum-mechanical states), and with respect to *that* state (whatever it is), the eigenstate–eigenvalue link must fail in this interpretation.

However, we must be quite careful about how we say that every observable has a value. We *can* say 'A has some definite value, and B has some definite value', and so on through all observables. This statement ends up being, in quantum logic, the meet of **1** with itself several times over, which is just **1** again. On the other hand, we must not say: 'the observables A, B, ... jointly have some set of definite values'. That statement is properly translated as

$$\bigvee_{i,j,\dots} \left(P^A_{a_i} \wedge P^B_{b_j} \wedge \cdots \right). \tag{4.3}$$

In lattice theory, the expression (4.3) is the zero subspace, which always has probability 0 and therefore corresponds to the always-false proposition. However, as I mentioned, the quantum logic interpretation as I use the term here says not that (4.3) is a contradiction, but that we cannot say it, or, that it is meaningless.

But *why*? Originally, Putnam — one of the main advocates of the quantum logic interpretation[2] — did say that (4.3) is utterable, but a contradiction. However, this original version of the quantum logic interpretation runs into several problems. Because it is committed to saying that the joint probability of $P^A_{a_i}, P^B_{b_j}, \dots$ is 0 for all $i, j, \dots$, it is also, apparently, committed to saying that the marginal probability for, say, $P^A_{a_i}$ will also be zero, for all a_i, and this consequence we know to be false. Something about Putnam's original quantum logic interpretation has to budge.

Later proponents of the quantum logic interpretation, including Putnam himself,[3] took the view that propositions involving non-orthogonal events are *undefined*, or *meaningless*. This move should look familiar — it was one way to avoid the Dutch Book argument against an epistemic interpretation of the quantum probability measure. Indeed, this move is what allows quantum logic to avoid the Dutch Book argument.

The quantum logic interpretation therefore proposes to solve the measurement problem with the simple postulate that quantum logic is the 'true' logic. Hence, for example, the statement 'for some i, the apparatus is in the state $|M_i\rangle$' is true, on this interpretation. Moreover, the quantum logic interpretation requires that propositions involving non-orthogonal events be meaningless. As the reader has already seen, probabilities on this interpretation (the single-time ones, at least) are epistemic.

4.1.2 The Kochen–Specker theorem and quantum logic

Thus far I have taken the set of all projection operators on a Hilbert space to be a lattice. However, as mentioned in chapter 1, they can also be considered to be a partial Boolean algebra. Quantum logic naturally endorses the latter. In a lattice, the operations meet, join, and orthocomplement are defined between all elements, but in a partial Boolean algebra, they are only defined for *compatible* elements, i.e., elements that are contained in some common Boolean subalgebra.

In other words, there is a (reflexive, symmetric, but not transitive) relation, c, defined on the set of all propositions. For any two propositions, P and Q, PcQ is read 'P is compatible with Q'. In a partial Boolean algebra, the lattice-theoretic operations are defined between P and Q if and only if PcQ. Moreover, PcQ if and only if P and Q are elements of some common Boolean subalgebra of the set of all propositions. Intuitively, PcQ if and only if P and Q are co-measurable. (Why? Because P and Q are elements of a common Boolean subalgebra if and only if they commute, $PQ - QP = 0$, and non-commuting observables cannot be simultaneously measured in standard quantum mechanics.)

However, the quantum logic interpretation must go well beyond the claim that statements involving non-orthogonal propositions are meaningless (or, undefined). The Kochen–Specker theorem forces the quantum logic interpretation to adopt a fairly radical perspectivalism, as I shall now describe.

In 1967, Kochen and Specker[4] proved that there is no map, $\mu : \mathscr{A} \to \{0, 1\}$, from the partial Boolean algebra, $\mathscr{A}$, of quantum propositions (on a Hilbert space of dimension greater than 2) to the Boolean algebra $\{0, 1\}$ that meets the following homomorphism condition: for all $P, Q \in \mathscr{A}$ such that PcQ

$$\mu(P \vee Q) = \mu(P) \vee \mu(Q), \tag{4.4}$$
$$\mu(P \wedge Q) = \mu(P) \wedge \mu(Q), \tag{4.5}$$
$$\mu(P^{\perp}) = \mu(P)^{\perp}. \tag{4.6}$$

Now, truth-valuations on $\mathscr{A}$ are just maps from $\mathscr{A}$ to $\{0, 1\}$. Hence the Kochen–Specker theorem shows that there is no homomorphic truth-valuation on $\mathscr{A}$. However, the quantum logic interpretation *does* require there to be truth-valuations on $\mathscr{A}$. Hence, the quantum logic interpretation must deny the homomorphism condition.

The question, obviously, is whether the homomorphism condition can be plausibly denied. Later in this chapter, I will suggest that modal interpretations *can* plausibly deny it, by denying that some observables have values at all. However, their way of denial is not open to the quantum logic interpretation, which assigns values to all observables.

The quantum logic interpretation does, however, have an explanation, of sorts, of the denial of the homorphism condition. Consider, for example, (4.4). If it holds, then $P \vee Q$ is true if and only if either P is true, or Q is true; but if quantum logic is the 'true' logic, then $P \vee Q$ is made true by *any* proposition contained in the subspace spanned by P and Q. In other words, in quantum logic, both P and Q can be false, while $P \vee Q$ (i.e., 'P or Q') is true. Hence, the supposition that the true logic of the world is quantum logic might be thought by itself to explain the failure of the homomorphism condition.

This line of reasoning is seductive, but it hides the truly radical length to which the quantum logic interpretation is forced by the Kochen–Specker theorem. To see why, consider the following condition, entailed by the homomorphism condition:

For each set, S, of mutually orthogonal elements of $\mathscr{A}$ that spans $\mathscr{H}$, μ maps one and only one element of S to 1, and the rest to 0.

The quantum logic interpretation *obeys* this condition, but then must deny whatever makes the difference between this condition and the homomorphism condition. In fact, the following is sufficient to get from the condition above to the homomorphism condition:

The map μ is single-valued: for every $P \in \mathscr{A}$, μ maps P to just one element of $\{0,1\}$.

This condition has been implicit in the discussion thus far, and is indeed usually presupposed of maps in general. However, the point is that one might assign *relativized* truth-values to the elements of $\mathscr{A}$. This condition rules out doing so.

In particular, it rules out the relativization of truth-assignments to Boolean subalgebras. Rather than seeking a truth-valuation over $\mathscr{A}$ in the form of a map from $\mathscr{A}$ to $\{0,1\}$, one might seek instead a set of truth-valuations, $\mu_{\mathscr{F}}$, one for each Boolean subalgebra, $\mathscr{F}$, of $\mathscr{A}$.

In these terms, the condition of single-valuedness may be restated as follows:

For any Boolean subalgebras, $\mathscr{F}_1$ and $\mathscr{F}_2$ of $\mathscr{A}$, and any P such that $P \in \mathscr{F}_1$ and $P \in \mathscr{F}_2$, $\mu_{\mathscr{F}_1}(P) = \mu_{\mathscr{F}_2}(P)$.

The *denial* of this condition leads to a radical form of perspectivalism: truth-valuations are relativized to Boolean subalgebras. I call this perspectivalism 'radical' because one and the same event (or proposition) can be 'true' when taken to be the member of *one* Boolean algebra, but 'false' when taken to be the member of another.

An example might help make the point clear. Consider a coarse-grained position-observable, A, with just three eigenspaces, each corresponding to some region of space: 'up' (eigenvalue $+1$), 'down' (eigenvalue -1), and 'middle' (eigenvalue 0). Suppose that for some system this observable has the value $+1$, so that the system in question can be said to occupy the region 'up'. Now consider a second observable, A', which shares with A the eigenspace corresponding to 'up', but divides up the space $P_{-1}^A \vee P_0^A$ differently, into the eigenspaces $P_{-1}^{A'}$ and $P_0^{A'}$. (That is, $P_i^{A'} \neq P_j^A$ for any $i, j = -1, 0$.) 'Relative to' this second observable, the particle will in general *not* have the value $+1$. That is, 'relative to' the first observable, the particle is in the region 'up', while 'relative to' the second observable, it is *not*.

Note, moreover, that the subspace $P_{-1}^A \vee P_0^A$ is in the Boolean algebra generated by the eigenspaces of A' (because we must have $P_{-1}^A \vee P_0^A = P_{-1}^{A'} \vee P_0^{A'}$), so that *not* having the property P_{+1}^A would seem to correspond in both cases to *having* the property of being in either the down region or the middle region. This claim, too, the quantum logic interpretation must deny, lest it end up saying that a particle both is, and is not, in the 'up' region. That is, a particle lacking the property $P_{+1}^{A'}$ must have the property $P_{-1}^{A'} \vee P_0^{A'}$. Hence the quantum logic interpretation must admit not only that truth-valuations are relativized to Boolean subalgebras, but also that the physical meaning of a given event in the algebra $\mathscr{A}$ is relativized to a Boolean subalgebra. Indeed, it is no longer clear that we are talking about the algebra $\mathscr{A}$ at all. After all, if one and the same element of $\mathscr{A}$ can have different truth-values (for the same system), and different physical interpretations, then why suppose that it is just a single element of an algebra? Instead, the quantum logic interpretation seems to be concerned with an unconnected set of Boolean algebras, one for each observable.

This feature of the quantum logic interpretation leads to my final point along these lines, namely, that the quantum logic interpretation must deny that the values of observables respect the functional relations among observables. Note that the map μ induces a map on observables, v, such that for any observable, A, $v[A]$ is the value of A. Kochen and Specker showed that the homomorphism condition is equivalent to the condition that for any (Borel) function, f, the map v obeys:

$$v[f(A)] = f(v[a]). \tag{4.7}$$

For example, if A has the value a, then A^2 should have the value a^2. The quantum logic interpretation must deny (4.7).

What is thus far lacking in the quantum logic interpretation is a clear account of why we should not be unhappy with these consequences. It is unsatisfactory to point out that they follow from the supposition that the logic of the world is quantum logic. What we need is an explanation of how it could be that (4.7) fails. After all, before we met the quantum logic interpretation, we might have been tempted to measure $f(A)$ by measuring A, then plugging the result into the function $f(\cdot)$. What we need is a physical account of why this procedure for measuring $f(A)$ is inadequate. More generally, what we need is a plausible argument in favor of perspectivalism.

4.1.3 Further challenges for the quantum logic interpretation

Aside from the challenge of making plausible the perspectivalism mentioned in the previous section, the quantum logic interpretation faces other challenges. In this section, I mention three of them, and discuss two of those three in detail.

The first challenge is dynamics. It is the avowed aim of most advocates of the quantum logic interpretation to be realists about the quantum world, but the attribution of definite properties at each time might not be sufficient for realism. Some might think it necessary that a *dynamics* of the possessed properties be given as well. Moreover, if quantum logic is to recapture the observed dynamics of macroscopic objects (and not just their observed non-dynamical properties), then apparently it must derive the dynamics of macroscopic objects from the dynamics of quantum objects. Thus far, advocates of the quantum logic interpretation have not proposed any dynamics. (The dynamics for modal interpretations offered later in this chapter could be modified, perhaps, to deliver a dynamics for the quantum logic interpretation. However, the problem of the non-existence of joint probabilities for non-orthogonal propositions would have to be overcome somehow.)

A second challenge is the measurement problem. Of course, advocates of the quantum logic interpretation will say that the measurment problem is solved in the strongest possible way: *every* observable has a definite value all of the time. The question is whether the sense in which that statement is true is the sense in which we believe it to be true. Consider, for example, what we say about the state of the measuring apparatus after a measurement: 'It will be $|M_1\rangle$, or $|M_2\rangle$, or, . . . '. Of course, what we *mean* by 'or' is the classical 'or'.

Therefore, the first task that the quantum logic interpretation faces, if it is to convince us that it solves the measurement problem, is to convince us that what we take to be a classical 'or' can be modelled by the lattice-theoretic 'join'. At first, this claim seems fairly plausible. After all, it often happens that science corrects our intuitions. The paradigmatic example in modern physics is in the theory of relativity, which is usually taken to show that our Euclidean intuitions about space and time are mistaken. Relativity theory replaces Euclidean geometry with a non-Euclidean geometry, and we have gotten used to it. Quantum logic wants to make a similar move for our classical logical intuitions.[5]

However, general relativity does not simply replace our old Euclidean intuitions with a non-Euclidean geometry. It also *explains* why we might have had those Euclidean intuitions in the first place: they are *approximately* correct, given the usual sorts of masses that we encounter in the region of space-time that we happen to occupy. It is primarily because general relativity provides an explanation of this sort that we can accept its prescription. Thus far, quantum logic has provided no analogous explanation. Classical logical intuition says, for example, that the distributive law holds: if 'p and (q or r)' is true, then so is '(p and q) or (p and r)'. In quantum logic, the distributive law does *not* hold. What quantum logic has thus far not explained is why most people *think* the distributive law holds, and indeed why it *does* hold, at least to very high accuracy, in the macroscopic world. (Explaining the latter might be sufficient for explaining the former.)

The most convincing explanation would be of the following form: Given the sorts of object that we happen to observe (i.e., that our perceptual apparatus can detect), the distributive law holds almost all of the time, or to a high degree of accuracy. Such an explanation would be forthcoming if the quantum logic interpretation could argue convincingly that the set of all properties detectable by us forms a Boolean algebra.

A third challenge facing the quantum logic interpretation is making plausible the claim that propositional combinations involving non-orthogonal events are meaningless, or undefined. Here too an explanation of the sort already offered for the failure of distributivity might ease the acceptance of this doctrine, but some at least might find the explanation less convincing in this case, for it must overcome the idea that we have private access to our own meanings. It *seems* to me that I know what I mean when I say 'the electron has momentum p and position q'. The quantum logic interpretation must convince me first that I *could* be mistaken, and second that I am.

These challenges are just that.[6] They are by no means mortal blows to the quantum logic interpretation, and indeed I have suggested at least some strategies for meeting them. Nonetheless, until they are met convincingly, we should think of them as, at least, potential problems. On the other hand, advocates of the quantum logic interpretation have given some positive arguments for their interpretation. Apart from the claim to solve the measurement problem, which I have already discussed, they have claimed to *derive* Lüder's rule from the other principles of the quantum logic interpretation. To finish my consideration of the quantum logic interpretation, I consider this claim.

The argument is due to Friedman and Putnam.[7] They consider only the case where the state of the system is a vector (i.e., a pure state), $|\psi\rangle$, in which case we may write Lüder's rule as:

$$p^{\psi}(P|P') = \frac{|\langle\psi|P'PP'|\psi\rangle|^2}{|\langle\psi|P'|\psi\rangle|^2}. \tag{4.8}$$

They assume that the state $|\psi\rangle$ is known, and they distinguish two cases, one in which P' is one-dimensional, and one in which it is not.

In the first, case, they say, imagine you are told that P' is true (or, occurrent). Then you will assign it probability 1. (Remember that the quantum logic interpretation takes quantum probabilities to be epistemic.) However, there is just *one* probability measure over $L_{\mathscr{H}}$ that assigns probability 1 to P' (when P' is one-dimensional). This fact follows immediately from Gleason's theorem, which says, recall, that the set of all probability measures over $L_{\mathscr{H}}$ is given by the set of all density opertors (including the pure states, of course) via (1.3). (See note 9.5 of chapter 1.) However, only one density operator (namely, the pure state P') assigns probability 1 to P'. Hence quantum logic plus Gleason's theorem is enough to get (4.8).

Before we consider the case where $\dim(P') > 1$, note that the argument given above used nothing about the quantum logic interpretation, apart from its adherence to the epistemic interpretation of the quantum probability measure. Any interpretation that agrees to the epistemic interpretation of the quantum probability measure can borrow this argument for (4.8).

However, the argument for the case where $\dim(P') > 1$ relies specifically on the principles of quantum logic. Friedman and Putnam say that two quantum propositions, P and Q, are equivalent (in quantum logic) if '(P and

Q) or (not-P and not-Q)' is true. Quantum-logically:

$$(P \wedge Q) \vee (P^{\perp} \wedge Q^{\perp}). \tag{4.9}$$

They then say that P and Q are equivalent *given* a quantum state, $|\psi\rangle$, if

$$P_{\psi} \leq \left[(P \wedge Q) \vee (P^{\perp} \wedge Q^{\perp})\right],$$

where P_{ψ} is the projection onto the subspace spanned by $|\psi\rangle$ ($= |\psi\rangle\langle\psi|$), and '$\leq$' is the relation for projections corresponding to 'is a subspace of' for subspaces. In particular, letting Q be the projection onto $P'|\psi\rangle$,

$$P_{\psi} \leq \left[(P' \wedge Q) \vee (P'^{\perp} \wedge Q^{\perp})\right]. \tag{4.10}$$

The equivalence between Q and P' given $|\psi\rangle$ seems to warrant an equivalence between $p^{\psi}(P|P')$ and $p^{\psi}(P|Q)$. The latter probability is already given by the first case, because Q is a one-dimensional projection. Hence we arrive again at (4.8).

There is one immediate problem with this argument, and it involves the statement that P_{ψ} implies the equivalence of P' and Q. Recall that P' was an arbitrary proposition, and that Q was the proposition corresponding to the projection of $|\psi\rangle$ onto P'. While the last two are always compatible (so that it is meaningful to assert propositional combinations of them),[8] neither of them is in general compatible with P_{ψ}. What, then, warrants our saying that P_{ψ} implies the equivalence of P' and Q? How is this statement meaningful?

One ultimately unsatisfactory answer is that statements of the form 'P implies Q' are not propositional combinations of P and Q. This claim goes well beyond any of the claims made by quantum logic up to now. Certainly it *seems* that 'P implies Q' is a proposition that combines P and Q. What story could the quantum logic interpretation tell to convince us otherwise? Following the earlier discussion, they may claim that we only believe that 'P implies Q' is a propositional combination of P and Q because it *is* so for the sorts of P and Q that we can observe. However, of the three times that the quantum logic interpretation has had to make such a claim, this time is the least convincing. It is very hard to see how the nature of the physical world could have anything to say about what counts as being a propositional combination. In any case, to make such a claim is to suppose that logic is empirical in a very strong sense indeed, and the analogy with geometry seems entirely to break down. Although the *properties* of geometric objects, triangles for example, are perhaps empirically determined, we do not normally take what *counts* as a geometrical object in the first place to be empirically determined.

Friedman and Putnam can easily revise their argument to avoid the need for this infelicitous definition of equivalence.[9] The revised argument begins by endorsing the classical definition of conditional probabilities whenever the events appearing in the conditional probability are compatible (i.e., commute). Then let $P' = Q + \tilde{Q}$, where $Q \perp \tilde{Q}$. (Recall that Q is the projection of $|\psi\rangle$ onto P'.) Then $p^{\psi}(Q|P') = 1$, for any P', because $p^{\psi}(\tilde{Q}) = 0$, so that

$$\frac{p^{\psi}(Q \wedge P')}{p^{\psi}(P')} = \frac{p^{\psi}(Q)}{p^{\psi}(Q + \tilde{Q})} = \frac{p^{\psi}(Q)}{p^{\psi}(Q)} = 1. \tag{4.11}$$

Then, given our knowledge that the 'original' state of the system is $|\psi\rangle$, and that P' has occurred, we must choose a revised ('conditionalized') probability measure that assigns probability 1 to Q. Because Q is one-dimensional, Gleason's theorem again entails that there is just one measure that assigns Q probability 1, namely, the one given by Lüder's rule.

This form of the argument is more satisfactory than the original, but still has some problems. I have already noted that it is restricted to pure quantum states, and in the real world, hardly *anything* (apart, presumably, from the universe) is in a pure state. Moreover, it is not clear how the argument would go for non-pure states, because it is not clear how to make a proposition out of a non-pure state. The most obvious way does not work. Call the subspace of vectors to which the state W assigns non-zero probability the 'kernel' of W, $K(W)$. Then the obvious way to generalize the argument is to let the proposition associated with W be $K(W)$. Letting $P_{K(W)}$ be the projection onto $K(W)$, and letting Q be the projection onto $K(P'WP')$, the relation needed for the second case is

$$P_{K(W)} \leq \left[(P' \wedge Q) \vee (P'^{\perp} \wedge Q^{\perp}) \right]. \tag{4.12}$$

This relation holds, but Q is in general not one-dimensional, so that we cannot piggy-back on the case of a one-dimensional P', as before.

Of course, the quantum logic interpretation can always use the argument mentioned in chapter 1, which relies on the condition (1.5). However, in this case, the quantum logic interpretation would again say nothing more than what âny epistemic interpretation of quantum probability measures would say. Indeed, if we rely on the revised version of the arugment in the case of an initial pure state (but multi-dimensional P'), then it is not even clear why any epistemic interpretation cannot avail itself of the entire argument. Hence, what began as an argument in favor of the quantum logic interpretation looks to have become an argument in favor of *any* epistemic interpretation.

A second problem involves a further claim made by Friedman and Putnam, namely, that the derivation of Lüder's rule provides an explanation of the phenomenon of interference in, for example, the two-slit experiment. In the two-slit experiment, an electron is fired at a screen with two slits, and then its position is measured on a screen some distance behind the slits. The probability of finding the particle in the region corresponding to the projection operator, P, is given by (2.6):

$$\frac{\mathsf{Tr}[U^{-1}(t')P'W(0)P'U(t')P]}{\mathsf{Tr}[W(0)P']}, \qquad (4.13)$$

where $P' = P_1 \vee P_2$, P_i is the projection associated with the particle's going through the ith slit, and $U(t')$ governs the evolution of the system between the time (time 0) when the particle passess through the slits, and the time (t') when the particle hits the screen.

However, we have already seen (section 2.2 that the move from Lüder's rule to (4.13) is non-trivial. Lüder's rule is a non-dynamical rule, while (4.13) makes a claim about dynamics. What the quantum logic interpretation needs is an account of dynamics that can justify the move from Lüder's rule to (4.13).

Such an account can, however, be given. It is already suggested by the alternative formulation of the quantum logic interpretation in terms of 'many quantum states', mentioned earlier. At each time, the most one can know about a quantnum system is a single pure state, $|\psi\rangle$. During periods in which one's knowledge is otherwise unchanged (by observation), how should we evolve the probability measure generated by $|\psi\rangle$? This question is evidently a question for physics rather than logic. It is analogous to the question 'given that a particle's classical position and momentum is (x, p) at time 0, and given that I make no observation between time 0 and time t, what should I say the position and momentum are at time t?' Of course, physics provides an answer to the question, in the form of the Schrödinger equation, which says how quantum states evolve. Hence quantum logic should say that, in the absence of knowledge gained otherwise, the Schrödinger equation governs the evolution of probability measures. This condition of the evolution of the values of observables is sufficient to underwrite Friedman and Putnam's claim that the quantum logic interpretation provides an explanation of interference phenomena.

These challenges should be taken as just that: challenges to an interpretation that should, like other interpretations, be taken seriously. These days, the quantum logic interpretation seems to be in some disrepute, but the problems raised here are really no more serious than the problems that other interpretations face.

4.2 Modal interpretations

4.2.1 General characterization of modal interpretations

Modal interpretations take the third of the three strategies that I outlined for avoiding the Dutch Book argument: they choose a subset of $L_{\mathscr{H}}$ to be the domain of the quantum probability measure. Given a judicious choice, it will turn out that the quantum probability measure *can* be interpreted epistemically, and that at the same time, the properties of everyday objects remain in the domain of the quantum probability measure.

All modal interpretations draw a distinction between theoretical states and physical states (different terms are used by different authors):

Physical state: A specification of all occurrent and non-occurrent events, or, all possessed properties, for a system.

Theoretical state: The state assigned by quantum mechanics. It determines all probability distributions over possible physical states.

In other words, the theoretical state yields probabilistic predictions about which events are now occurrent, and which will be occurrent in the future. (Note, however, that the theoretical state alone does not provide a dynamical picture of how the occurrent events evolve in time. Many dynamical schemes will be compatible with a given set of probability distributions at different times.)

Different modal interpretations define the set of possible physical states differently, and they do so by defining a set of 'definite-valued events'. In every modal interpretation, this set depends at least in part on the theoretical state, and may be denoted '$\mathscr{A}_W$'. The set of possible physical states is a set of maps from $\mathscr{A}_W$ to the set {'does not occur', 'occurs'}, or $\{0, 1\}$. Finally, modal interpretations define a probability measure over the possible physical states, by adopting the quantum probability measure over $\mathscr{A}_W$.

By denying the eigenspace–eigenvalue link, modal interpretations have a straightforward way out of the measurement problem: Even if the theoretical state assigns probability 1 to a superposition of macroscopically distinct states, modal interpretations may say that the physical state specifies that only one term of the superposition corresponds to an occurrent event. The simple way to be certain that they may tell this story about *every* such superposition is to take $\mathscr{A}_W$ to be $L_{\mathscr{H}}$, but this option is not open to modal interpretations.

One problem with taking $\mathscr{A}_W$ to be $L_{\mathscr{H}}$ should be clear from chapter 1: there do not exist joint probability measures over arbitrary elements of $L_{\mathscr{H}}$, and hence there does not exist a probability measure over the maps from

$L_{\mathscr{H}}$ to (the Boolean algebra) $\{0, 1\}$, for such a measure would induce a joint probability measure over arbitrary elements of $L_{\mathscr{H}}$.

However, there is a more fundamental problem: there do not even exist homomorphisms from $L_{\mathscr{H}}$ to $\{0, 1\}$ in the first place. As we saw in the previous section, the reason lies in the Kochen–Specker theorem. Modal interpretations do, in general, accept that truth-valuations on $\mathscr{A}_W$ should be homomorphisms from $\mathscr{A}_W$ to $\{0, 1\}$, so that the Kochen–Specker theorem does apply. Note that in accepting this requirement that truth-valuations be homomorphisms, modal interpretations respect in a weaker way the dictum of the quantum logic interpretation that the operations of join, meet, and orthocomplement are the logical operations. Modal interpretations require that truth-valuations on $\mathscr{A}_W$ respect the quantum-logical operations. Hence modal interpretations must choose a proper *subset* of $L_{\mathscr{H}}$ to be the set, $\mathscr{A}_W$, of definite-valued properties for a system in the state W.

Many modal interpretations choose a 'faux-Boolean algebra', and in the next section, I define these structures and show that choosing such a structure is sufficient to satisfy certain 'classicality' constraints on the probabilities. In section 4.2.3, we will see how to use these classicality constraints, together with some other conditions, to motivate some particular versions of the modal interpretation. In section 4.2.4 I consider an objection to modal interpretations. In section 4.2.5, I discuss some problems and questions that arise in the consideration of compound systems and the structure of properties generally. In section 4.2.6, I discuss whether modal interpretations solve the measurement problem. Finally, in section 4.2.7, I argue that to avoid the projection postulate, modal interpretations must find a dynamics. A framework for a dynamics has indeed been developed, and I review it briefly. I emphasize that, as with my previous discussions, my discussion of modal interpretations is relatively brief. There is a large and growing literature on modal interpretations and there are a number of open questions. Most of the literature, and much of the recent research, I will merely summarize or skip.

4.2.2 Faux-Boolean algebras

In each modal interpretation discussed here, $\mathscr{A}_W$ turns out to be a *faux-Boolean algebra*.[10] In this section I define and state some facts about faux-Boolean algebras that will help to motivate the modal interpretations of subsequent sections.

A faux-Boolean algebra is one that is constructible as follows. Take any set, S, of mutually orthogonal projections in some Hilbert space, $\mathscr{H}$. Let

$S^\perp$ be the set of all projections orthogonal to the span of the elements of S. Take the union of S and $S^\perp$, and close under the operations join, meet, and orthocomplement — denote the result by $\mathsf{close}(S \cup S^\perp)$. The algebra $\mathsf{close}(S \cup S^\perp)$ is a faux-Boolean algebra.[11]

Finally, define a 'faux-classical measure' over a faux-Boolean algebra, $\mathscr{B}$, to be any measure over $\mathscr{B}$ that assigns probability 0 to everything in $S^\perp$. Given such a measure, p_{fc}, the ordered triple $\langle S \cup S^\perp, \mathsf{close}(S \cup S^\perp), p_{\text{fc}} \rangle$ may be called a 'faux-classical' probability theory. (The set $S \cup S^\perp$ plays the role of a sample space, because the elements in $S \cup S^\perp$ are the atoms of $\mathsf{close}(S \cup S^\perp)$.)

Faux-classical probability theories are so called because, although the algebra of events is a non-Boolean algebra, they have a classical model, in a sense that I shall now make precise. Given some (possibly non-Boolean) algebra of events, $\mathscr{C}$, marginal probabilities, $\mathsf{prob}[P]$, for all $P \in \mathscr{C}$, and conditional probabilities, $\mathsf{prob}[P_1|P_2]$, for all $P_1, P_2 \in \mathscr{C}$, a *classical model for* $\mathscr{C}$, $\mathsf{prob}[\,\cdot\,]$ *and* $\mathsf{prob}[\,\cdot\,|\,\cdot\,]$ is given by:

(i) a (classical) probability space $\langle \Omega, \mathscr{F}, \mathsf{p} \rangle$, where Ω is a sample space, $\mathscr{F}$ the Boolean σ-algebra generated by Ω, and p a classical probability measure; and

(ii) a map, $\mu : \mathscr{C} \to \mathscr{F}$, such that for any $P_1, P_2 \in \mathscr{C}$

$$\text{(a)} \quad \mathsf{prob}[P_1] = \mathsf{p}\big[\mu[P_1]\big]$$

$$\text{(b)} \quad \mathsf{prob}[P_1|P_2] = \frac{\mathsf{p}\big[\mu[P_1] \cap \mu[P_2]\big]}{\mathsf{p}\big[\mu[P_2]\big]}.$$

Note that μ is not required to be a morphism of any type, and that it is allowed to be many-to-one and not onto. There are stricter definitions of the existence of a classical model, for example one that required μ to be one-to-one, or a homomorphism, or an isomorphism, but only the weaker definition is needed here.

Given only an algebra of events, $\mathscr{C}$, we may say that $\mathscr{C}$ admits a classical model if and only if there are some probability measures $\mathsf{prob}[\,\cdot\,]$, and $\mathsf{prob}[\,\cdot\,|\,\cdot\,]$ defined over all of $\mathscr{C}$, such that $\mathscr{C}$ together with these measures admits a classical model.

We may now state the precise sense in which faux-classical probability theories have classical models.[12]

Theorem 4.1 *Given any faux-classical probability theory,* $\langle S \cup S^\perp, \mathsf{close}(S \cup S^\perp), p_{\text{fc}} \rangle$, *where* p_{fc} *is generated by some density operator,* W, *and letting the con-*

ditional probabilties, p_{fccond}, on close($S \cup S^{\perp}$) be defined by Lüder's rule, the triple $\langle$close($S \cup S^{\perp}$), $p_{\text{fc}}, p_{\text{fccond}}\rangle$ has a classical model.

The basic idea behind the proof of this theorem is simple. The 'Boolean part' of a faux-Boolean algebra (i.e., the part generated by S alone) behaves classically, and so of course admits a classical model — just map it isomorphically to some classical algebra, $\mathscr{F}$. The 'null part' (i.e., the part generated by $S^{\perp}$ alone) is in general non-classical, but by choosing a faux-classical measure, and mapping the null part to the zero element of $\mathscr{F}$, all non-classicality is 'hidden', because the entire non-classical part of the faux-Boolean algebra has probability 0. (The only (minor) complication in the proof arises from the fact that faux-Boolean algebras also have events formed from the join of an event from the Boolean part with an event from the null part.)

Perhaps this strategy for making a non-classical algebra appear classical sounds artificial, but in fact there are some natural faux-Boolean algebras defined in quantum-mechanical terms and some natural faux-classical measures over them. In the next three subsections, we shall see three such algebras.

The next two theorems provide further evidence for faux-Boolean algebras' being interesting structures. The first is actually quite obvious, given the Kochen–Specker theorem, though it happens to hold for Hilbert spaces of any dimension greater than 1 (whereas, recall, the Kochen–Specker theorem holds only for spaces of dimension greater than 2):[13]

Theorem 4.2 *Given any Hilbert space $\mathscr{H}$ (dim($\mathscr{H}$) > 1) and any pure state W on $\mathscr{H}$, $L_{\mathscr{H}}$, together with $p^{W}[\,\cdot\,]$ (defined for all elements of $L_{\mathscr{H}}$) and $p^{W}[\,\cdot\mid\cdot\,]$ (defined via Lüder's rule for all pairs of elements of $L_{\mathscr{H}}$), does not admit a classical model.*

The third theorem is again about faux-Boolean algebras and classical models. It uses the notion of a 'maximal' faux-Boolean algebra, which is one for which all of the atoms of the Boolean part (and therefore all of the atoms) are one-dimensional.

Theorem 4.3 *Given any Hilbert space $\mathscr{H}$ (dim($\mathscr{H}$) > 1), and any maximal faux-Boolean algebra, $\mathscr{B}$, consider any projection $P \notin \mathscr{B}$, $P \in \mathscr{H}$. Then for any faux-classical measure on $\mathscr{B}$, which is generated by some density operator, W, the triple $\langle$close($\mathscr{B} \cup P$), $p^{W}[\,\cdot\,], p^{W}[\,\cdot\mid\cdot\,]\rangle$ does not admit a classical model.*

Theorem 4.3 indicates that maximal faux-Boolean algebras are 'maximally classical'. It does not say that some 'larger' structures (in some sense of 'larger') might not have classical models, but if one is for any reason committed to the definiteness of a maximal faux-Boolean algebra and wants a classical model, then theorem 4.3 shows that nothing more can be made definite.

In general, then, the importance of these results lies in how they affect what we choose for $\mathscr{A}_W$. Theorem 4.1 tells us that any $\mathscr{A}_W$ that is a faux-Boolean algebra will admit a classical model. (It remains to show, however, that p^W is a faux-classical measure on $\mathscr{A}_W$.) Theorem 4.2 appears to place only a weak restriction on what we may choose $\mathscr{A}_W$ to be, if we want $\mathscr{A}_W$ to have a classical model that respects the measures over $\mathscr{A}_W$ given by $p^W[\,\cdot\,]$ and $p^W[\,\cdot\,|\,\cdot\,]$. However, in conjunction with other conditions, theorem 4.2 plays an important role in motivating various types of modal interpretation. The role of theorem 4.3 is usually to show that the algebra $\mathscr{A}_W$ as chosen by a given modal interpretation is maximally classical, relative to other conditions. In the next subsections, I state some of the results that these theorems may be used to prove.

4.2.3 Motivating modal interpretations

4.2.3.1 Van Fraassen's 'Copenhagen variant'

Van Fraassen's 'Copenhagen variant' of the modal interpretation[14] has in common with all other modal interpretations the distinction between theoretical and physical states.[15] Van Fraassen defines $\mathscr{A}_W$ in terms of a 'representative' of the physical state. Given a theoretical state, W, he says that the representative of the physical state is *some* pure state, $|\psi\rangle$, lying in the support of W; $|\psi\rangle$ represents the physical state in the sense that an event, P, is occurrent if and only if $p^\psi(P) = 1$. Van Fraassen does not give an explicit condition for an event's being non-occurrent, and one might want to add the condition that an event is non-occurrent if and only if $p^\psi(P) = 0$. I discuss this addition below.

As van Fraassen is aware, this prescription alone is insufficient, for the set of possible representatives does not yield a set of definite-valued events over which a classical probability measure can be defined. It will not do to say that $\mathscr{A}_W$ is the set of all P such that $p^\psi(P) = 1$ (or 0) for *some* $|\psi\rangle$ in the support of W. If the support of W is of dimension greater than 2, then this set does not admit any (homomorphic) truth-valuations, because of the Kochen–Specker theorem. Nor does it admit a classical model (even if the dimension of the support of W is 2).

True to Copenhagen and to his own 'constructive empiricism', van Fraassen turns to measurements. He does not claim to discuss *all* types of measurement, but focuses on what he calls 'von Neumann–Lüders' measurements,[16] in which the value of an apparatus-observable, M, becomes correlated with the value of a measured observable, A. Van Fraassen further distinguishes non-disturbing measurements (where the measured system ends up in the eigenstate of A indicated by M) from disturbing measurements (where the measured system does not end up in the state indicated by M).

In the light of this account of measurement, van Fraassen adopts the following rule:[17] At the end of a non-disturbing von Neumann–Lüders measurement, the possible representatives of the physical state are all $P_{a_i}^A \otimes P_{m_i}^M$ with non-zero probability. (The probability that a projection $P_{a_i}^A \otimes P_{m_i}^M$ is the representative is just the usual quantum-mechanical probability for the occurrence of $P_{a_i} \otimes P_{m_i}$ at the end of the measurement.) If the measurement is disturbing, then the rule applies only to the pointer observable, M (i.e., possible representatives are the $P_{m_i}^M$ with non-zero probability). Outside of measurement situations, the representative of a system in the theoretical state W is just the projection onto the support of W.

Let us consider what the structure of $\mathscr{A}_W$ is in van Fraassen's interpretation. There are three possibilities. First, $\mathscr{A}_W$ is the Boolean algebra generated by all $P_{a_i}^A \otimes P_{m_i}^M$ with non-zero probability. Second, $\mathscr{A}_W$ is the Boolean algebra generated by all $P_{a_i}^A \otimes P_{m_i}^M$ with non-zero probability, plus the subspace orthogonal to their span. Third, $\mathscr{A}_W$ is the Boolean algebra generated by all $P_{a_i}^A \otimes P_{m_i}^M$, plus all projections in the subspace orthogonal to their span. The third possibility is the most liberal, and it is the one that I shall assume henceforth, though one might have reasons for preferring one of the others. Call this algebra '$\mathscr{A}_{vF}$'.

$\mathscr{A}_{vF}$ (in any of its versions given above) is easily seen to be a faux-Boolean algebra. Moreover, $\mathscr{A}_{vF}$, together with all of the quantum-mechanical probabilities for measurement-outcomes, admits a classical model. Indeed, it is useful to consider van Fraassen's interpretation in terms of the following four conditions. (I give the conditions here in terms of non-disturbing measurements, but they are easily modified to cover disturbing measurements, in the way noted above, and all of the arguments remain valid.)

Copenhagen condition: After a non-disturbing von Neumann-Lüders measurement on a system in the state W, the projections $P_{a_i}^A \otimes P_{m_i}^M$ with non-zero probability are atoms of $\mathscr{A}_W$ (where A is the measured observable, and M is the apparatus-observable).

The motivation for this condition is, presumably, some form of empiricism, which I shall not discuss here.

The next condition is:

Null space condition: $\mathscr{A}_W$ contains every projection in the space orthogonal to the space spanned by all of the $P_{a_i}^A \otimes P_{m_i}^M$ with non-zero probability.

The motivation here is clear. If an event definitely cannot occur (because it has probability 0), then we would like to say that the event is non-occurrent.

The next condition is:

Classicality condition: $\mathscr{A}_W$, together with $p^W[\,\cdot\,]$ and $p^W[\,\cdot\mid\cdot\,]$, admits a classical model.

The classicality condition can recognize that quantum mechanics is non-classical. Indeed, in general faux-Boolean algebras are non-Boolean. The point of adopting the classicality condition is not so much to return to a classical world-view as to provide for the possibility of interpreting probabilities epistemically. This possibility in turn makes solving the measurement problem easier, because a probability distribution over the indicator-states of an apparatus can then be interpreted epistemically: the apparatus possesses one of the indicator-states, and the probabilities represent our ignorance about which state is possessed.

The last restriction on $\mathscr{A}_W$ is that it form an algebra (more precisely, an ortholattice). The intuitive reason is that, for example, if the events P and P' are possible properties, then 'P or P'' should also be a possible property. Hence:

Closure condition: $\mathscr{A}_W$ is closed under countable meet, join, and orthocomplement.

One might wonder why closure uses lattice-theoretic operations rather than classical set-theoretic operations. The reason is not that the former are more intuitive, or natural, than the latter, but that the latter simply do not make sense for subspaces. For example, $P \cup P'$ is not in general a subspace, even if P and P' are. Note, however, that closure does not commit one to a quantum-logical view, for we still require that $\mathscr{A}_W$ admit a classical model. (There are other objections to the closure condition, particularly from those who prefer partial Boolean algebras over ortholattices, but I shall not pause to consider them here.)

Given these conditions, we may state the following:

Theorem 4.4 *The Copenhagen, null space, classicality, and closure conditions are together equivalent to the choice of $\mathscr{A}_W$ as $\mathscr{A}_{vF}$.*

The proof follows almost immediately from theorem 4.3, and may be given here. The Copenhagen, null space, and closure conditions are sufficient to generate $\mathscr{A}_{vF}$. Now, although the atoms of the Boolean part of $\mathscr{A}_{vF}$ are not in general one-dimensional, they are required, by the Copenhagen condition, to be atoms. Therefore no subspace of any of them can be definite-valued, and no definite-valued event can intersect one of them (because then the intersection would also be definite-valued, by closure). Therefore, the atoms of the Boolean part of $\mathscr{A}_{vF}$ may be treated as one-dimensional. Then by theorem 4.3, nothing can be added to $\mathscr{A}_{vF}$ without violating the classicality condition.

Because of the Copenhagen condition, van Fraassen's account is indeed very Copenhagen-like. In particular, van Fraassen's account is silent about the possible properties of a system outside of measurement-situations. (Of course, it remains unclear exactly what counts as a measurement. It might be that systems are almost always being 'measured', due to their being correlated with the environment, for example.) This consequence might be unpalatable to the convicted realist, but an empiricist like van Fraassen need not be worried.

However, van Fraassen's neglect of a dynamics for the possessed properties of a system between measurements leads to another problem. Modal interpretations propose to deny the projection postulate. *Prima facie*, however, it is not clear how they can do so plausibly. After all, the quantum-mechanical state of a system is used in modal interpretations to generate an epistemic probability measure, but then when one of the events over which the measure is defined *occurs* (or, is known to occur), surely the measure should be changed to reflect this fact. In other words, surely the quantum-mechanical state should be collapsed.[18]

Now, most modal interpretations have an answer to this argument. They can point out that the quantum-mechanical state not only determines probabilities, but also plays a crucial role in the dynamics of possessed properties. (I will discuss the dynamics of modal interpretations later.) Hence the quantum-mechanical state has something to do besides determine probabilities. (In this respect, modal interpretations resemble Bohm's theory closely. I will discuss Bohm's theory in the next chapter.) However, van Fraassen's Copenhagen variant is pointedly agnostic about dynamics: the quantum state at the end of a measurement therefore plays just one role, that of determining (epistemic) probabilities. It is therefore unclear why we should *not* apply the projection postulate in this case.

If the Copenhagen variant *does* include the projection postulate, however, then what exactly is its value as an interpretation? How, for example, does it go beyond the 'standard interpretation', for which van Fraassen himself has few kind words?[19] The Copenhagen variant suggests some potentially precise criteria for when the projection postulate is to be applied. For example, when the state of an observer becomes correlated 'in the right way' to the value of an observable, A, on a system, then that observer attributes a collapsed theoretical state to the system, i.e., an eigenstate of A. The only imprecision here is in the phrase 'the right way'. I insert it because probably there are ways of being correlated with the value of A on a system that would *not* lead one to attribute a collapsed state to the system. Spelling out what counts as being correlated 'in the right way' would involve spelling out, in physical detail, what counts as 'observation of a value', and that is a project for empiricists that I gladly leave to them. Another possible criterion is that collapse occurs upon measurement, i.e., as soon as $\mathscr{A}_W$ is defined, though this criterion relies on the imprecise notion of 'measurement'.

In any case, assuming that the Copenhagen variant can answer the question of when to apply the projection postulate, there remains the question of whether it succeeds in describing our classical experience. Because $\mathscr{A}_{vF}$ is a faux-Boolean algebra, it at least succeeds in making our experience conform to the rules of classical probability theory. That success, however, is not enough. Will we ever witness a superposition of macroscopically distinct states on this view? In order to answer 'no', one would have to argue that the act of human observation is, formally at least, an act of measurement. However, even that is not enough, for it is possible in principle to measure observables whose eigenspaces represent macroscopically 'smeared' states of the sort we never see. Therefore, one would have to argue further that the 'measurements' that constitute human observation are never of such observables. I will not conjecture here whether such an argument could be made convincingly.

4.2.3.2 The Kochen–Dieks–Healey modal interpretation

The modal interpretations of Kochen, Dieks, and Healey differ in some important ways,[20] but I consider them together under the heading of a single modal interpretation, as formulated below.[21]

To give the formulation, recall that every density operator, W, has a unique spectral resolution, $\{P_i\}$, such that

$$W = \sum_i w_i P_i, \tag{4.14}$$

where no two of the w_i are equal, and none of the w_i is 0. Such a decomposition always exists and is always unique, though some of the P_i may be multi-dimensional. Finally, note that because the P_i are mutually orthogonal, they generate a Boolean algebra, which we may denote $\mathscr{B}_W$. The algebra,

$\mathscr{A}_{\text{KDH}}$, of definite properties in the Kochen–Dieks–Healey interpretation is

$$\mathscr{A}_{\text{KDH}} = \{P \,|\, P = \bar{P} \vee \tilde{P} \;\text{ for some }\; \bar{P} \perp (\bigvee_i P_i) \;\text{ and some }\; \tilde{P} \in \mathscr{B}_W\}. \quad (4.15)$$

Clifton has shown that $\mathscr{A}_{\text{KDH}}$ thus defined includes $\mathscr{A}_W$ as given by Kochen, Dieks, and Healey in the original formulations of their interpretations,[22] which were in terms of the biorthogonal decomposition theorem:

Theorem 4.5 (biorthogonal decomposition theorem) *Given a vector $|\psi\rangle$ in the tensor-product Hilbert space $\mathscr{H}^\alpha \otimes \mathscr{H}^\beta$, $|\psi\rangle$ can always be written*

$$|\psi\rangle = \sum_i c_i |u_i\rangle \otimes |v_i\rangle, \quad (4.16)$$

where $\{|u_i\rangle\}$ is an orthonormal basis in $\mathscr{H}^\alpha$ and $\{|v_i\rangle\}$ is an orthonormal basis in $\mathscr{H}^\beta$. (Some of the c_i can be zero.) Moreover, this decomposition is unique whenever $|c_i| = |c_j|$ implies $i = j$.[23]

Given (4.16), $\mathscr{A}_{|\psi\rangle\langle\psi|}$ as generated by the original formulations of this interpretation is the Boolean algebra generated by $\{|u_i\rangle\langle u_i| \otimes |v_i\rangle\langle v_i|\}$. The probability measure over $\mathscr{A}_W$ is the quantum-mechanical measure.

Arntzenius has noted that this original proposal fails to meet a purportedly natural condition on definite-valued events, namely, that every event having probability 0 or 1 is definite-valued.[24] Clifton's formulation using faux-Boolean algebras resolves this difficulty, as well as the difficulty of how to define $\mathscr{A}_W$ when the biorthogonal decomposition is not unique.

The probability measure over $\mathscr{A}_{\text{KDH}}$ is, of course, the quantum-mechanical one, p^W (where W is the state used to generate $\mathscr{A}_{\text{KDH}}$). In addition, the Kochen–Dieks–Healey interpretation provides joint probabilities for the simultaneous occurrence of events for two or more different subsystems of a given composite system. Let the state of a composite system be W, a density operator on the tensor-product Hilbert space $\mathscr{H}^\alpha \otimes \mathscr{H}^\beta$. Let the states of the two subsystems be W^α and W^β (obtained by a partial trace), and let the elements in the spectral resolutions of W^α and W^β be P_i^α, and P_j^β respectively. Then the joint probability that the pair (P_i^α, P_j^β) occurs is $\text{Tr}[W P_i^\alpha \otimes P_j^\beta]$. This prescription can be extended to more than two subsystems in the obvious way.[25]

The definition of $\mathscr{A}_{\text{KDH}}$ can be motivated by some physically intuitive conditions. Clifton has shown that the Kochen–Dieks–Healey interpretation implies and is implied by two sets of such conditions.[26] I shall not review his theorems here, but instead give a similar motivation, based on slightly different conditions.

Like the Copenhagen variant, the Kochen–Dieks–Healey interpretation assumes the closure condition and the classicality condition. It also assumes the null space condition, but as a consequence of this stronger condition:

Certainty condition: For any P, if $\text{Tr}[WP] = 0$ or 1 then $P \in \mathscr{A}_W$.

The certainty condition may be motivated by the supposition that if the result of a measurement can be predicted with certainty, then the corresponding event is occurrent. However, this condition is not undeniable — once we have given up the eigenspace–eigenvalue rule it is no longer *obvious* that events with probability 1 are occurrent.

The last restriction on $\mathscr{A}_W$ is unique to the Kochen–Dieks–Healey interpretation — actually, it is motivated by a condition of Clifton's[27] — and applies to mixed states only. It is motivated by the intuition that if $W = \sum_i w_i P_i$, then the system *could* be in some pure state corresponding to any one-dimensional $P \subseteq P_i$, for any P_i. However, it is well known that to go beyond this statement invites either contradiction or arbitrariness, because there is no unique decomposition of a projection, P_i, whose dimension is greater than 1, into one-dimensional projections. The best one can say is that *some* one-dimensional subspace of *some* P_i might be occurrent, though in general we cannot say which, nor even specify some subset of such projections as the possibly occurrent ones. Given this claim about the meaning of mixed states, we require:

Weak ignorance condition: For each P_i in the spectral resolution of W and each one-dimensional $P \leq P_i$, $\mathscr{A}_W \subseteq \mathscr{A}_P$.

The intuition behind the weak ignorance condition is that the discovery that some one-dimensional subspace of some P_i in the spectral resolution of W is occurrent should only increase (or at most, leave unchanged) what we know to be definite-valued. This motivation is perhaps less convincing in the case of so-called 'improper' mixtures (i.e., mixed states obtained by tracing out part of a compound system whose components are entangled), but, as Clifton notes,[28] proper and improper mixtures are formally indistinguishable, and because the weak ignorance condition is reasonable for proper mixtures, it must also hold, at least formally, for improper mixtures. (Clifton's version of the weak ignorance condition is weaker than the one here, however.)

We may now state the following:

Theorem 4.6 *The closure, certainty, weak ignorance, and classicality conditions are together equivalent to the definition of $\mathscr{A}_W$ as $\mathscr{A}_{\text{KDH}}$.*

The proof of this theorem is omitted here.[29]

Theorem 4.6 shows a nice physical motivation for the Kochen–Dieks–Healey interpretation, and it shows that in this interpretation our experience will at least obey the rules of classical probability theory. Again, however, the question arises: Does it permit the observation of superpositions of macroscopically distinct states? This question may be divided into two parts. First, does the Kochen–Dieks–Healey interpretation at least make apparatuses indicate a definite value after a measurement? Second, does the Kochen–Dieks–Healey interpretation permit humans to observe superpositions of macroscopically distinct states?

The answer to the first question is not obvious. The problem was raised by Albert.[30] He argued that modal interpretations fail to assign definite indicator-states to apparatuses in the case of non-ideal measurements. (It is obvious that the Kochen–Dieks–Healey modal interpretations give a satisfactory account of ideal measurements — one can see this point most easily by considering the formulation in terms of the biorthogonal decomposition thoerem.) In reply, it was argued that the Kochen–Dieks–Healey modal interpretations can solve the problems that arise from non-ideal measurements.[31] In particular, it was pointed out that in the case of non-ideal measurements, those interpretations assign properties that are very 'close' (in Hilbert space norm) to the ideal indicator-states. And, as I argued in the case of CSL, 'close' is good enough.

However, these arguments are all in the context of apparatus-observables with discrete spectra. It has also been argued[32] that the Kochen–Dieks–Healey interpretation might find it difficult, if not impossible, to make the same sort of argument in the context of observables with continuous spectra. However, the discussion is complicated by the fact that it is unclear even how to *formulate* this interpretation in that case.

The answer to the second question is equally unclear, though there are perhaps more possibilities for arguing 'no' than there are in the Copenhagen variant. If the argument I suggested for the Copenhagen variant can be made — i.e., if human observation can be properly modelled as the measurement of observables with eigenspaces that do not represent superpositions of macroscopically distinct states — then the Kochen–Dieks–Healey interpretation can use that argument, if it can account for measurements. However, if that argument fails, then there are perhaps other ways to solve the problem within the Kochen–Dieks–Healey interpretation. On the other hand, it has been argued[33] that under some realistic situations, the modal interpretation seems to attribute entirely unsatisfactory states — states that are even maximally far (in Hilbert space norm) from well-localized states — to macroscopic objects. There remain some possibilities for avoiding this consequence, but it is clear that the Kochen–Dieks–Healey modal interpretations have a great deal of work to do on this front.

4.2.3.3 Bub's modal interpretation

Bub's modal interpretation begins with the intuition that some events, or events of a certain kind, are *always* definite-valued. Further, it is always possible to take the eigenspaces of (at least) one operator as being in $\mathscr{A}_W$ while still meeting the classicality condition and avoiding a Kochen–Specker contradiction (because the eigenspaces of one operator generate a Boolean algebra). Bub therefore proposes to take some observable, R, as always having a definite value. He then asks: How large can we make $\mathscr{A}_W$ while still avoiding a Kochen–Specker contradiction, and requiring that $\mathscr{A}_W$ be a lattice (perhaps non-Boolean)? Such lattices Bub calls 'maximal determinate sublattices', denoted here by '$\mathscr{A}_{Bub}$'.

When the spectral resolution of W is a single projection ($W = P/\text{Tr}[P]$), $\mathscr{A}_{Bub}$ is constructed as a faux-Boolean algebra, taking the 'generating' set, S to be $\{(P \vee P_{r_i}^{\perp}) \wedge P_{r_i}\}$, each element of which is the projection of P onto the eigenspace, P_{r_i} ($= P_{r_i}^R$). When the spectral resolution, $\{P_i\}$, of W has more than one element, the maximal determinate sublattice for each P_i is determined, as if the state were $P_i/\text{Tr}[P_i]$, and $\mathscr{A}_{Bub}$ is the intersection of all these lattices.

Like the others, Bub's interpretation can be motivated by a set of physically meaningful conditions. These include the closure condition and the classicality condition, plus two conditions that are unique to Bub's approach. The first is:

Definite-observable condition: For some observable, R, all of the eigenspaces of R are in $\mathscr{A}_W$.

The second is:

Null projection condition: If W is a pure state, P, then every projection in the span of all $P_{r_i}^R \perp P$ is in $\mathscr{A}_W$.

The definite-observable condition is what best characterizes Bub's interpretation. The null projection condition is a little less satisfactory, given Bub's overall approach (of taking R to be fundamental). Why should projections that are incompatible with R be in $\mathscr{A}_W$? Bub can get away with the null projection condition because it adds only projections with probability 0, thus maintaining the possibility of a classical model, but the question here is why one would *want* to get away with the null projection condition? Bub's own motivation[34] seems to be just that he wants to make $\mathscr{A}_W$ as large as possible without generating a Kochen–Specker contradiction. However, while that is an interesting mathematical exercise, it is not clear what it adds to an interpretation.

In any case, I shall not modify his interpretation here, for it does at least have a motivation, as the others do, in terms of the given conditions:

Theorem 4.7 *If W is pure, then the closure, classicality, null projection, and definite-observable conditions are together equivalent to the definition of* $\mathscr{A}_W$ *as* $\mathscr{A}_{\mathrm{Bub}}$.

The proof is trivial enough to give here. The definite-observable, null projection, and closure conditions are sufficient to generate $\mathscr{A}_{\mathrm{Bub}}$. The result is a maximal faux-Boolean algebra. Therefore, by classicality and theorem 4.3, nothing further can be added to $\mathscr{A}_R$.

There is an apparent weakness in this theorem, namely, that it applies only to pure states. Although I believe that theorems similar to theorem 4.7 hold in the general case, I do not have any such theorem to hand.[35] On the other hand, if we consider the interpretation to be applied first of all to the entire universe, then presumably the limitation to pure states is fine — the universe is plausibly assumed to be in a pure state.

Again, we have a physical motivation for $\mathscr{A}_{\mathrm{Bub}}$, and again we are assured that $\mathscr{A}_{\mathrm{Bub}}$ guarantees that our classical experience will obey the rules of classical probability theory, but again we must ask: Does it permit the observation of superpositions of macroscopically distinct states? Here, finally, the answer depends on a relatively simple issue: Does R have any eigenspaces that correspond to superpositions of macroscopically distinct states? Perhaps one can always *choose* R so that it does not. In that case, it appears that Bub's interpretation will not violate our classical experience. This feature of Bub's interpretation gives it an advantage over other modal interpretations, though some might consider the advantage to be won at the cost of the choice of R being *ad hoc*.

4.2.4 'Naïve realism' about operators

It has been pointed out[36] that the the measurement problem may be a manifestation of a quite fundamental error, which is labelled 'naïve realism about operators'. The error in question is to suppose that there is a one-to-one correspondence between operators (more precisely, self-adjoint operators on a Hilbert space) and physical quantities, and that this one-to-one correspondence is fixed for all time.

Such a supposition can lead to all sorts of trouble, for it encourages one to forget that in general it is the physical situation that determines how a given operator represents reality (or whether it does at all). Consider, for example, a standard Stern–Gerlach experiment (a measurement of the spin of a spin-1/2 particle — see chapter 6 for a more detailed account of such measurements). If one measures spin in the z-direction, then a certain region on the detector is associated with spin-up and some other region is associated with spin-down, so that each of the relevant eigenspaces of the spin-z observable, σ_z, is associated with a certain physical event, namely, a flash of the detector in a given region. Now consider the same measurement, but with an apparatus whose magnets are reversed. (Figure 6.1 should make these points clear.) In that case, the very same observable, σ_z is associated with physical events in a different way. In fact, the regions associated with the eigenspaces of σ_z are reversed — the region that *was* associated with the eigenspace corresponding to positive spin in the z-direction is now associated with the eigenspace corresponding to negative spin in the z-direction, and vice versa. To put the point more generally, the physical situation influences how operators (or eigenspaces of operators) correpond to physical events.

It is easy to forget this fact, and doing so can lead rather quickly to the measurement problem. However, even an interpretation that purports to *solve* the measurement problem may be guilty of naïve realism about operators, and modal interpretations, because they place such importance on eigenspaces of observables, might be thought to commit this particular sin.

In fact, however, they do not. As I will explain in detail in section 4.2.5.2, modal interpretations can agree to the proposition that the physical situation determines the role that an operator plays, i.e., how it represents reality (or whether it does so at all).

However, there is a sin apparently related to the sin of ignoring how the physical situation determines the role of operators, which is failing to *begin* one's interpretation with a physical intuition, and only later developing a formalism to express this intuition. It is somtimes claimed that Bohm's theory follows this line of development — as we will see in chapter 5, the central intuition there is that the world consists of particles in motion. Only after this intuition is expressed, the story goes, is the formalism to describe it developed.

Of course, this view oversimplifies the relation between physical intuition and formalism, but it might nonetheless express an important point, namely, that formalism devoid of physical meaning is no good as an interpretation of quantum mechanics. Indeed, it is difficult to disagree with this claim.

However, it is easy to take this idea too far, and doing so would lead to another (specious) objection to modal interpretations.

The objection is that modal interpretations do not begin with a detailed physical intuition about the way the world is. There are at least two reasons for rejecting this objection to modal interpretations.

First, it appears to assume that *nothing* interesting or important can be said about the physical world without the prior exposition of a clear physical intuition, but surely this requirement is too strong a restriction on theorizing about the physical world. After all, it seems at least possible that some quite general notions about physical reality could be expressed outside of a specific physical intuition (such as that driving Bohm's theory). Modal interpretations, at least, are predicated on the idea that such general notions can be expressed.

Second, modal interpretations *can*, as we have seen, be motivated by physically meaningful conditions, albeit conditions of a fairly general sort. Now, one may object that the conditions in question are physically mean-ingful only if one already assumes naïve realism about operators, but this objection fails in the face of the conditions actually used (for example, in the previous section). For example, the definite observable condition used to motivate Bub's interpretation is completely compatible with the idea that the physical situation determines how operators represent physical reality. From this point of view, Bub's interpretation, for example, expresses what can be said *prior to* a detailed exposition of some physical intuition, one that would, presumably, lead to the choice of some particular observable as definite-valued.

4.2.5 *Compound systems and the structure of properties*

4.2.5.1 *Options for the treatment of compound systems*

The previous subsections focused on single systems, but modal interpreta-tions differ not only in how they select $\mathscr{A}_W$, but also in their treatment of compound systems. The basic differences concern four points: the defi-nition of a subsystem, property composition, property decomposition, and supervenience. A detailed account of the position of each existing modal in-terpretation on these points is out of the question. Below is a brief discussion, with a few examples.

Consider a compound system composed of two subsystems, α and β. The Hilbert space for the compound system is $\mathscr{H}^{\alpha\beta} = \mathscr{H}^{\alpha} \otimes \mathscr{H}^{\beta}$. A generic property for α (projection onto $\mathscr{H}^{\alpha}$) is denoted P^{α}, and likewise for P^{β}.

The problem of defining a subsystem arises by considering a second factorization of $\mathcal{H}^{\alpha\beta}$ into $\mathcal{H}^{\alpha'}$ and $\mathcal{H}^{\beta'}$. (For the sake of emphasizing the problem, suppose that the dimensions of the primed and unprimed spaces are the same, so that the primed factorization is just a 'rotation' of the unprimed one.) We may assume that subsystems α and β receive their own algebras of definite properties according to the procedure of some modal interpretation. The question now is whether α' and β' do also. Not all advocates of modal interpretations have considered this question, but among those who have, Dieks has taken the view that every factorization determines a subsystem,[37] while Healey takes the view that there is a 'preferred' factorization that determines what the subsystems are.

Property composition is the condition that if α possesses P^α and β possesses P^β, then the compound system $\alpha\&\beta$ possesses $P^\alpha \otimes P^\beta$. Property decomposition is the condition that if $\alpha\&\beta$ possesses $P^\alpha \otimes P^\beta$, then α possesses P^α and β possesses P^β. Supervenience is the condition that a compound system possess only properties that are products of properties of subsystems or algebraic combinations of such properties. That is, supposing that $\{P_i^\alpha\}$ and $\{P_j^\beta\}$ are the definite properties for α and β, $\alpha\&\beta$ can have as definite properties only those in the lattice-theoretic closure of $\{P_i^\alpha \otimes P_j^\beta\}$. Some or all of these conditions are denied by various modal interpretations. (Note that the interpretations as given earlier concerned single systems only; they are all free to accept or deny any of the conditions given above.)

For example, as already mentioned, the discussion of Bub and Clifton suggests that an algebra of definite properties is given first of all for the universe (presumed to be in a pure state). Properties are then assigned to subsystems through the condition that P^α is definite for α if and only if $P^\alpha \otimes 1^\beta$ (a projection on the Hilbert space of the universe) is definite for the universe. (They do not say explicitly whether every factorization represents a subsystem.) Property composition, decomposition, and supervenience appear to hold in Bub's interpretation.

Healey imposes a number of conditions on algebras of events for subsystems, but it remains unclear to me what the consequences of these conditions are for Healey's set of definite properties. It appears that this set supports property composition and property decomposition, but not supervenience.

Vermaas and Dieks appear to deny all three.[38] They treat every system on its own, deriving an algebra of definite properties from the system's density operator. Consider, for example, the density operator $W^{\alpha\beta}$, and the reduced density operators W^α and W^β (obtained from $W^{\alpha\beta}$ by partial tracing). Clearly the eigenprojections of $W^{\alpha\beta}$ need not be tensor products of eigenprojections of W^α and W^β. Therefore, because the compound system

$\alpha\&\beta$ gets all of its properties by applying Vermaas and Dieks' procedure to $W^{\alpha\beta}$, property composition fails. Perhaps less obviously, property decomposition fails as well. To see why, consider a case where $\alpha\&\beta$ has possible properties $P_1^\alpha \otimes P_1^\beta$ and $P_2^\alpha \otimes P_2^\beta$, and where $P_1^\beta \perp P_2^\beta$ but $P_1^\alpha \not\perp P_2^\alpha$. Then α will not have either P_1^α or P_2^α as possessed properties, so that property decomposition must fail. Supervenience clearly fails as well.

As a final example, Bacciagaluppi and Dickson have proposed an interpretation according to which faux-Boolean algebras are assigned (along the lines advocated by Vermaas and Dieks) only to a chosen set of 'atomic' systems (given by some preferred factorization); all properties of compound systems are derived from the properties of the atomic systems via property composition (and algebraic combinations of such properties).[39] In this proposal, property composition of course holds, as do property decomposition and supervenience.

Although each interpretation is free to choose how to treat compound systems, the choices are not without consequences. It has been shown that the proposal to treat *every* factorization as defining a genuine subsystem leads to a Kochen–Specker contradiction. Apparently, the only way around this contradiction is to adopt some form of contextuality.[40] But this result ought not really worry anybody. The idea of a preferred factorization is not, perhaps, as *ad hoc* as it might at first appear. After all, assuming that the universe is really made up of, say, electrons, quarks, and so on, it makes good sense to take these objects to be the 'real' constituents of the universe, i.e., the bearers of properties that do not necessarily supervene on the properties of subsystems. Indeed, it would appear strange, if not downright silly, to suppose that, for example, a 'system' composed of the spatial degrees of freedom of some electron and the spin degrees of freedom of some atom is a genuine subsystem of the universe, deserving of its own properties (apart from those properties that it inherits by virtue of its being composed of two other systems).

However, even proposals adopting a preferred factorization can face severe constraints. Consider, for example, a system whose Hilbert space is $\mathscr{H}^{\alpha\beta} = \mathscr{H}^\alpha \otimes \mathscr{H}^\beta$, and one of its subsystems, α, whose Hilbert space is $\mathscr{H}^\alpha$. It can easily happen that a spectral projection, $P^{\alpha\beta}$, of $W^{\alpha\beta}$ (the quantum-mechanical state of the compound system $\alpha\&\beta$) does not commute with $P^\alpha \otimes 1^\beta$ (where P^α is a spectral projection of W^α). However, it is not obvious that we can define a joint probability for $\alpha\&\beta$ to possess $P^{\alpha\beta}$ and α to possess P^α. Indeed, as we have seen, there is in general no expression for the joint probability for non-commuting projections that is valid in every quantum-mechanical state. (Whether or not we wish to say that $\alpha\&\beta$ possesses $P^\alpha \otimes 1^\beta$

whenever α possesses P^{α}, a joint probability for $\alpha \& \beta$ to possess $P^{\alpha\beta}$ and α to possess P^{α} will induce a joint probability measure for $P^{\alpha\beta}$ and $P^{\alpha} \otimes 1^{\beta}$.)

On the other hand, modal interpretations do not need a general expression for the joint probability of any two non-commuting projections, but only expressions that are valid for limited sets of non-commuting projections, and in a limited number of quantum-mechanical states. Nevertheless, while some have tried to find such expressions, and have succeeded in special cases, no generally acceptable expression has yet been found. Indeed, some results suggest that no satisfactory expression will be found.[41]

Finally, note that the result of some of the choices made by some modal interpretations is that the final algebra of definite properties for some systems will not be a faux-Boolean algebra. In the case of Dieks' interpretation and Vermaas and Dieks' generalization,[42] this fact is obvious, because they assign properties to a subsystem based on any factorization of the Hilbert space for the universe of which it is a part. In the case of Healey, the conditions imposed appear to result in a set of definite properties that is not even a partial Boolean algebra,[43] much less a faux-Boolean algebra. On the other hand, the interpretation of Bub and Clifton and the proposal of Bacciagaluppi and Dickson yield algebras for any subsystem that is a faux-Boolean algebra.

4.2.5.2 The logical structure of properties

Having chosen some set, $\mathscr{A}_W$, of definite-valued events, modal interpretations as I introduced them say that all and only the propositions in $\mathscr{A}_W$ have a definite truth-value. (Any event, P, can be considered also to represent the proposition 'P occurs'.) We might well wonder why, especially when, in different circumstances, i.e., had $\mathscr{A}_W$ been different, different propositions would have had truth-values. Consider, for example, a spin-1/2 particle, with a two-dimensional Hilbert space (ignoring its spatial degrees of freedom). It may happen that $\mathscr{A}_W$ is determined by letting S be $\{P_{z,+}, P_{z,-}\}$ (where $P_{z,\pm}$ is the projection onto the ± 1 eigenstate of z-spin). In that case, $\mathscr{A}_W$ does not contain $P_{x,+}$. Are we to say that the proposition (expressed by the sentence) 'the system has positive spin in the x-direction' has no truth-value? In the case of Bohm's theory (as I will discuss in the next chapter), such a statement is plausible because it can be given a physical explanation — the conditions required for the proposition 'the system has positive spin in the z-direction' to be possibly true are physically incompatible with the conditions required for 'the system has positive spin in the x-direction' to be possibly true. However, the modal interpretation has no such explanation, at least not yet. Even in the case of Bohm's theory, I am tempted to say not that 'the system has positive spin in the x-direction' is neither true nor false, in the context of a measurement of z-spin, but that it is simply false.

In case you do not share my intuition, consider an everyday example. Suppose you say, 'I've got something in mind; guess its properties.' I say, 'I guess that it is green'. No matter what you have got in mind, and no matter what its actual properties are, it seems wrong to suppose that my guess is neither right nor wrong. You might, for example, have in mind Beethoven's 7th symphony — which is not even the sort of thing (let us suppose) that can be green — but even then it seems wrong to say that my guess is neither right nor wrong. Hardly more plausible is the claim that the sentence 'it is green' really expresses the proposition 'it is the sort of thing that can be green, and it is green'. The business of second-guessing the meaning of a sentence to satisfy tentative philosophic hypotheses is shady indeed. Rather, it seems one should say that what you have in mind is not green. My guess was wrong.

However, there is a difference between the sense in which Beethoven's 7th is not green, and the sense in which it is not in C minor. This difference might be thought to be captured by the proposal to adopt a three-valued logic, in which the truth-values are 'true', 'false', and 'middle',[44] so that 'Beethoven's 7th is in C minor' is false, while 'Beethoven's 7th is green' is middle. I shall not consider this approach here, however, but instead construct a language, $\mathscr{A}$, and a logic for modal interpretations that is classical and two-valued and nonetheless avoids the problem.

Let the connectives in $\mathscr{A}$ be the usual classical (Boolean) ones, denoted '&' (and), '∘' (or), and '¬' (not); the set of propositions in $\mathscr{A}$ is constructed by first putting in every proposition represented by a subspace in Hilbert space, then closing under the (classical) logical operators. I write $\mathscr{A} = \langle A, \&, \circ, \neg \rangle$. The language $\mathscr{A}$ contains many propositions not representable as subspaces of Hilbert space. Indeed, what we have done is to embed the usual quantum-propositional structure into a classical one.

What about the Kochen–Specker theorem? Did it not show that such an embedding does not exist? Not quite. What it showed is that there does not exist a truth-assignment from $L_{\mathscr{H}}$ to the Boolean algebra $\{0, 1\}$ that preserves the lattice-theoretic operations. It follows that the map from $L_{\mathscr{H}}$ to $\{0, 1\}$ induced by any map from $\mathscr{A}$ to $\{0, 1\}$ fails to preserve the lattice-theoretic operations. Moreover, there are at least two good reasons to preserve them, even though we are abandoning them as our logical operators.

First, they work empirically, and it would appear that the classical operators do not. Consider the two-slit experiment. There, we want to say that the particle passed through one slit or the other, but if we represent this disjunction classically, we make incorrect predictions. By representing it lattice-theoretically, we make correct predictions. Second, denying the homomorphism condition has its own bizarre consequences. For example, it may turn out that a single observable on a single system will have many values, rather than just one. Surely this consequence is not worth whatever gain the embedding will bring.

Both of these objections can be answered. Concerning the first, we will find that the lattice-theoretic operators are equivalent to the classical ones, when restricted to the empirically accessible propositions. Hence modal interpretations with the non-homomorphic embedding (non-homomorphic with respect to the lattice-theoretic operations) given above can account for the empirical success of the lattice-theoretic operations.

Concerning the second objection, it is true that denial of the homomorphism condition can have strange consequences, but it need not. To see why, recast the example above, using a 'color' observable and a 'key' observable. Beethoven's 7th clearly has a value for the 'key' observable, but it appears to lack any value for the 'color' observable. In violation of the homomorphism condition, then, we may say that some observables have no value (i.e., all eigenspaces are mapped to 0), but no observable has more than one value.

This situation is arguably not uncommon in physics, even classical physics. For example, one might say that a gas in a chamber has no surface tension, or that oak has no melting point. Classical systems can apparently lack values for some classical observables.

Hence the lattice of subspaces of Hilbert space can be embedded into a classical logic. More precisely, a classical truth-assignment — a map to the Boolean algebra $\{0, 1\}$ — is possible because we may deny that every mutually orthogonal set of subspaces must have one and only one element mapped to 1; some observables have every eigenspace mapped to 0.

Hence I propose that the language of modal interpretations should be based on $\mathscr{A}$. (I say 'based on $\mathscr{A}$' because I will add a modal operator to $\mathscr{A}$ below.) The semantics of $\mathscr{A}$ is the next order of business.

A truth-assignment on $\mathscr{A}$ begins with the specification of a faux-Boolean algebra, $\mathscr{B}$ (as given by some modal interpretation), and some subspace, P, in the set S used to generate $\mathscr{B}$; P is presumed true. Every proposition in $\mathscr{A}$ that is also in $\mathscr{B}$ is true if and only if it contains P. Every proposition in $L_{\mathscr{H}}$ but not in $\mathscr{B}$ is false. This prescription covers all propositions in $\mathscr{A}$ that are not generated from others by the classical connectives. The truth-value of all other propositions in $\mathscr{A}$ is determined in the usual, classical, truth-functional way. Denote the resulting map from $\mathscr{A}$ to $\{0, 1\}$ by $\tau_{\mathscr{B},P} : \mathscr{A} \rightarrow \{0, 1\}$.

Every $\tau_{\mathcal{B},P}$ induces a truth-assignment, $\tau_{\mathcal{B},P}|_{L_{\mathcal{H}}}$ on $L_{\mathcal{H}}$. It has been shown[45] that for every such $\tau_{\mathcal{B},P}|_{L_{\mathcal{H}}}$, the restriction of $\tau_{\mathcal{B},P}|_{L_{\mathcal{H}}}$ to $\mathcal{B}$, is homomorphic with regard to the lattice-theoretic operations. Because the elements of $\mathcal{B}$ are the only empirically available propositions (according to modal interpretations), we may conclude that $L_{\mathcal{H}}$ and $\mathcal{A}$ are empirically equivalent. Moreover, as suggested earlier, the denial of the homomorphism condition for propositions outside $\mathcal{B}$ is plausible when all such propositions are false. Now I will consider a logical foundation for this claim.

The central idea is that propositions outside a system's faux-Boolean algebra (as chosen by some modal interpretation) are necessarily false.[46] It is conceivable that plausible physical considerations could underlie this claim. For example, we do not simply postulate that a gas in a chamber does not have a surface tension. Reflection on the physical conditions required for a system of a certain kind (e.g., a gas in a chamber) to have properties of a certain kind (e.g., some surface tension) leads us to say that some things are physically incapable of having certain properties. Similarly, the fact that modal interpretations select a proper subset of the propositions in $\mathcal{A}$ as 'physically possible' might be understandable through physical considerations. The faux-Boolean algebra of propositions (or perhaps better, the set S that generates it) for a given system might be seen as characterizing 'what sort of thing' the system is, and it seems plausible that, given such a characterization, there might well be properties that that sort of thing cannot possess, by virtue of its being that sort of thing. Or, if that way of describing the impossibility sounds too essentialist, we may say that the complex of properties that the system has by virtue of having a definite truth-value for everything in $\mathcal{B}$ is physically incompatible with having properties in $L_{\mathcal{H}}$ but outside $\mathcal{B}$. (It remains an open question — hence an interesting challenge to modal interpretations — whether a credible account of this incompatibility can be given.)

The necessity operator being considered here is to be interpreted as representing neither logical necessity nor physical necessity. Rather, it represents 'necessity relative to a given physical situation'. For example, we do not say that it is physically impossible for the gas in the chamber to have a surface tension. The object to which we refer — ultimately, the particles that consitute the gas — might well have a surface tension under different circumstances. However, relative to the physical situation, it cannot have a surface tension, by virtue of an incompatibility between the conditions necessary for having a surface tension and the physical situation of the object. Similarly, we may suppose that a quantum-mechanical system's 'situation' is given by, or in any case is somehow connected with, its faux-Boolean algebra, and that this situation renders the possession of properties outside the faux-Boolean algebra 'impossible'.

Hence, for example, there remains a sense in which a spin-1/2 particle whose faux-Boolean algebra contains neither $P_{x,+}$ nor $P_{x,-}$ nonetheless could have spin-up in the x-direction. However, according to the present proposal, there is also a sense of 'necessity' according to which, relative to the situation of the particle (defined by its faux-Boolean algebra), the particle cannot have spin-up in the x-direction.

To implement this proposal, make the following definitions. First, a world, w, is given by a triple, $\langle \mathscr{B}, S, P \rangle$ (S must be specified explicitly because when $\mathscr{B}$ is Boolean, knowing $\mathscr{B}$ is not sufficient for knowing S). I assume that a world characterizes just a single system (though this 'single' system could be the entire universe). Second, the relation of accessibility [47] among worlds is such that $w' = \langle \mathscr{B}', S', P' \rangle$ is possible relative to w if and only if $S' = S$. These definitions imply that every proposition in $\mathscr{B}$ and every proposition in $\mathscr{A}$ entailed by some proposition in $\mathscr{B}$ is possibly true at $\langle \mathscr{B}, S, P \rangle$.

We now have a complete modal propositional language, $\mathscr{A}_{\text{modal}}$, a complete semantics given by a set of worlds, each of which generates a truth-assignment on $\mathscr{A}$, and an accessibility relation between worlds which we use to get truth-assignments on all of $\mathscr{A}_{\text{modal}}$.

I have already said that the logic proposed here is empirically equivalent to standard quantum logic. I can now make that statement more precise. For every world, $w = \langle \mathscr{B}, S, P \rangle$, there is no possibility that, for any $P_1, P_2 \in \mathscr{B}$, any of $P_1 \vee P_2$, $P_1 \wedge P2$, or $P_1^{\perp}$ differs in truth-value from $P_1 \circ P_2$, $P_1 \& P_2$, or $\neg P_1$, respectively. (Note that although $\vee$, $\wedge$, and $^{\perp}$ are not symbols in the language $\mathscr{A}$, they nonetheless can be used to express propositions in $\mathscr{A}$, because every proposition in $L_{\mathscr{H}}$ is in $\mathscr{A}$. Hence $P_1 \vee P_2$ above should be read: 'the proposition in $\mathscr{A}$ corresponding to the proposition $P_1 \vee P_2$ in $L_{\mathscr{H}}$,' and so on.)

This notion of empirical equivalence is underwritten in part by the impossibility of propositions that are represented by subspaces not in $\mathscr{B}$ — there is no need to match logical structure on these propositions, because it is not possible to discover the discrepancy. In other words, the truth-assignment $\tau_{\mathscr{B}, P}|_{L_{\mathscr{H}}}$ need not be homomorphic (with regard to the lattice-theoretic operations) except when restricted to $\mathscr{B}$.

One might object that some of the new propositions in $\mathscr{A}$ (those not in $L_{\mathscr{H}}$) *are* empirically accessible but not in $\mathscr{B}$. For example, the proposition $P \circ Q$, where P is in $\mathscr{B}$ and Q is not, is empirically accessible — it can be empirically known to be true, if P is — but $P \circ Q$ is not in $\mathscr{B}$. This objection prompts a distinction between 'directly empirically accessible' and 'indirectly empirically accessible'. All possibly-true propositions that are not in $\mathscr{B}$ can be true only by virtue of some proposition in $\mathscr{B}$ being true. We may therefore call these 'indirectly empirically accessible'. They can be empirically known to be true, but only because some proposition in $\mathscr{B}$ is empirically known to be true. Only the propositions in $\mathscr{B}$ are directly empirically accessible, and all of our empirical knowledge rests on our knowledge about them. In other words, we can test an indirectly accessible proposition only by testing a directly accessible one. Hence equivalence of the classical and lattice-theoretic operators on $\mathscr{B}$ is all that is needed to show that as far as testing empirically accessible propositions is concerned, there is no difference between the two.

The conclusion, then, is that the non-classicality of quantum logic may be purely phenomenological. Given that some observables may lack values some of the time, there is the possibility of a completely classical 'subquantum' logic. Modal interpretations therefore force us to rethink the claim that the logic of quantum theory is necessarily non-classical.

4.2.6 Dynamics

4.2.6.1 Is a dynamics needed?

I have noted that modal interpretations provide, for every time t, the set of possible modal states, $S(t)$, and a probability measure over this set. However, one wants to know more. I take it to be crucial for any modal interpretation (or, at least, any modal interpretation that denies the projection postulate — see section 4.2.3.1) that it also answer questions of the form: Given that a system possesses property P at time s, what is the probability that it will possess property P' at time t ($t > s$)? In other words, we need a *dynamics* of possessed properties.

It is evident that in the Kochen–Dieks–Healey approach, at least, the dynamics we are after will be genuinely probabilistic. This point can be seen in a trivial example. Let

$$|\Psi(t)\rangle = \sum_i c_i(t)|\alpha_i\rangle \otimes |\beta_i\rangle, \tag{4.17}$$

where $|\alpha_i\rangle \in \mathscr{H}^\alpha$ and $|\beta_i\rangle \in \mathscr{H}^\beta$. Then the spectral resolution of $W^\alpha(t)$ is at all times given by $\{P_i^\alpha\}$ $(= \{|\alpha_i\rangle\langle\alpha_i|\})$, unless there happens to be a degeneracy (in which case the projections in the unique spectral resolution of W^α would be given by sums of elements of $\{P_i^\alpha\}$) — and I will assume for this example that there are no degeneracies. However, the probabilities attached to these spectral projections will change, due to the time-dependence of the $c_i(t)$.

Now, consider an ensemble of systems, all with statevector $|\Psi(t)\rangle$ and with modal states distributed across $\{P_i^\alpha\}$ according to the quantum-mechanical probability measure, $|c_i(t)|^2$. As the distribution, $|c_i(t)|^2$, changes in time, some members of the ensemble must make transitions among the $\{P_i^\alpha\}$ in order to preserve the distribution.

Why can these transitions not be deterministic? Because, assuming that the modal interpretation is 'complete' (i.e., assuming that the modal state specifies completely the physical state of the system), there can be nothing to distinguish those systems that make a transition from P_1^α to P_2^α (for example) from those systems that make a transition from P_1^α to some property other than P_2^α (or make no transition at all). There is therefore nothing in the theory that could have 'determined' the transition from P_1^α to P_2^α.

The same argument can be applied to Bub's interpretation. It is clear that, given some preferred observable, one can find a state and an evolution for that state that will generate a set, $S(t)$, at least some of whose elements are constant, but whose probabilities change. (However, this argument breaks down when we move to observables with continuous spectra.[48]) What we need, then, is a stochastic dynamics.

Or do we? Some may consider a dynamics of possessed properties to be superfluous. After all, could quantum mechanics not get away with just single-time probabilities? Why can we not settle for an interpretation that supplements standard quantum mechanics *only* by providing in a systematic way a set (the set of possible properties) over which its single-time probabilities are defined? If we require of this set that it include the everyday properties of macroscopic objects, then what more do we need?

What we need is an assurance that the *trajectories* of possessed properties are, at least for macroscopic objects, more or less as we see them to be. For example, we should require not only that the book at rest on the desk have a definite location, but also that, if undisturbed, its location relative to the desk does not change in time. Hence one cannot get away with simply specifying the definite properties at each time. We need also to be shown that this specification is at least *compatible* with a reasonable dynamics. Even better, we would like to see the dynamics explicitly.

Of course, modal interpretations do admit — trivially — an 'unreasonable' dynamics, namely, one in which there is no correlation from one time to the next. (In this case, probability of a transition from the property P at s to P' at t is just the single-time probability for P' at t.) In such a case, the book on the table might *not* remain at rest relative to the table, even if undisturbed. I take it that such dynamics are not very interesting and fail to provide any assurance that modal interpretations can describe the world more or less as we think it is.

Without assurance of the existence of a reasonable dynamics, then, modal interpretations are far less attractive than they might otherwise be. My main task here is to give some indication of how a dynamics can be constructed for modal interpretations. This work was done jointly with Guido Bacciagaluppi.[49]

4.2.6.2 *How to make a dynamics*

To begin, we need to establish a one-to-one correspondence between elements of $S(t)$ and $S(t')$ for any two times, t and t'. ($S(t)$ is the set used to generate a system's faux-Boolean algebra of definite properties at time t.) That is, we need a time-indexed family of one-to-one maps, $\kappa_t : S(t) \to S(t + dt)$. These maps will not themselves give us the dynamics, but will merely be used in establishing the dynamics.

Immediately one fears that the $S(t)$ may have different cardinality for different t, thus ruling out the existence of the κ_t. There is really no problem here, however. We can simply find the $S(t)$ with the greatest cardinality. Then decompose the elements of $S(t)$ at the other times into 'fiduciary' elements. For example, suppose that the set of greatest cardinality has cardinality N, while another set, at time s, has cardinality $N - 2$. In that case, there must be some multi-dimensional projections in either $S(s)$ or $S^{\perp}(s)$ (because the dimension of a system's Hilbert space does not change in time). Suppose for illustration that $S(s)$ contains a three-dimensional element, P. Then we can decompose P into three one-dimensional, mutually orthogonal, projections, $P = \tilde{P}_1 + \tilde{P}_2 + \tilde{P}_3$, replacing P in $S(s)$ with these three projections. The resulting faux-Boolean algebra will contain the old one as a subalgebra. Hence any dynamics involving the new algebra will induce a dynamics on the old algebra. (For example, if any of the $\tilde{P}_i$ is 'possessed', then we will say that P is possessed.) Hence we can always arrange things so that the maps κ_t exist.[50]

Of course, one could then simply *choose* a family of such maps, and doing so would be sufficient to move on to the next step. However, much better would be a principled method to determine which family of maps one should use. Under some circumstances at least, there is such a method. In Bub's interpretation, this point is obvious. If $P_i(t)$ is the projection of the system's statevector onto the ith eigenprojection of the preferred observable, then simply choose the map (at t) that maps $P_i(t)$ to $P_i(t + dt)$. In the Kochen-Dieks-Healey version, it is less trivial to find such a map, but it has been shown that under fairly plausible circumstances, the elements of the spectral resolution of any system's reduced state (density operator) evolve continuously.[51] Hence, again, the map that takes each element of $S(t)$ to its continuous evolute in $S(t + dt)$ is clearly the most natural map to choose.

Henceforth I shall consider just the Kochen–Dieks–Healey interpretation, though everything said about that interpretation can easily be translated into corresponding statements about Bub's interpretation, and indeed to *any* interpretation that chooses at every time a faux-Boolean algebra to represent the possible properties of a system.

The existence of the family of maps κ_t can be used to 'transform away' the time-dependence of the elements of $S(t)$, thus greatly simplifying the problem of finding a dynamics. To see why, consider two extreme cases. In one, the system evolves freely, so that the eigenvalues of its reduced density operator do not change, while the eigenprojections evolve unitarily. In the other, the eigenvalues change, but the eigenprojections do not. In the general case, both eigenvalues and eigenprojections evolve non-trivially. However, the one-to-one correspondence allows us to consider only extreme cases of the second type. Every general case can be mapped onto such an extreme case by letting each eigenprojection be mapped at every time to the (unique) eigenprojection at $t = 0$ from which it evolved. This mapping carries the eigenvalue with it, so that we have a case in which the eigenprojections do not evolve, but the eigenvalues change just as they did in the general case. Hence from now on we may consider just the second extreme case.

In general, one should do so for a composite system, $\alpha\&\beta\&\cdots$ whose state, W, is a state on the tensor-product Hilbert space $\mathscr{H} = \mathscr{H}^\alpha \otimes \mathscr{H}^\beta \otimes \cdots$, and whose reduced states are W^α, W^β, and so on. However, to simplify the notation, I shall let a *single* index range over the composite properties $P_k^\alpha \otimes P_l^\beta \otimes \cdots$. I let the index i range over these composite properties at time t and j range over them at time $t + dt$, so that the probability of a transition from composite property i (at time t) to composite property j (at time $t + dt$) may be written simply as $p_{j|i}$. Having defined the $p_{j|i}$, the joint transition probabilities, one may then readily obtain transition probabilities for subsystems (including single systems) by taking marginals.

To follow this procedure generally, one would need joint probabilities defined over all subsystems to which one assigns properties using the rules of the Kochen–Dieks–Healey interpretation. Hence in Dieks' approach, one needs a joint probability for all systems at every level of the hierarchy in every factorization. In Healey's approach, one needs a joint probability for all systems at every level of a single preferred factorization. In the atomic approach, one needs joint probabilties only for the atomic systems in a single preferred factorization. Of course, as we have seen, only the last joint probabilities are known to exist. Hence the procedure here is known to apply only to the atomic version (i.e., the proposal of Bacciagaluppi and Dickson — see section 4.2.5.1).

There are several constraints that an expression for $p_{j|i}$ should meet. First, it should recover the single-time joint probabilities:

$$\sum_i p_{j|i} p_i = p_j. \tag{4.18}$$

If it meets this requirement, then automatically it also gets the single-time marginal probabilities for the subsystems right.

A second requirement is that $p_{j|i}$ satisfy a master equation. To formulate this requirement, let $T_{j|i}$ be a matrix such that $T_{j|i}dt$ is the probability of a joint transition from the (compound) state i to the (compound) state j in the interval of time dt. Also, let $J_{j|i}$ be the joint probability current between eigenprojections. That is, J is the 'net flow' of probability:

$$J_{i|j} = T_{i|j} p_j - T_{j|i} p_i, \tag{4.19}$$

so that J is clearly anti-symmetric. Moreover, it is clear from (4.19) that

$$T_{i|j} = \frac{T_{j|i} p_i - J_{j|i}}{p_j}, \tag{4.20}$$

a fact that I will use below. (Note that $T_{j|i}$ and $J_{j|i}$ are time-dependent quantities, but I shall usually leave the time-dependence implicit.) Given its definition (and the conservation of probability), $J_{j|i}$ must satisfy the continuity equation:

$$\dot{p}_j = \sum_i J_{j|i}. \tag{4.21}$$

Finally, using (4.21) and (4.20), we see that $T_{j|i}$ must obey a master equation:

$$\dot{p}_j = \sum_i \left(T_{j|i} p_i - T_{i|j} p_j \right). \tag{4.22}$$

As Vink has noted,[52] given $J_{j|i}$ and p_j satisfying (4.21), the master equation (4.22) is satisfied by any $T_{j|i}$ satisfying

$$J_{j|i} = T_{j|i} p_i - T_{i|j} p_j, \tag{4.23}$$

where $T_{j|i} \geq 0$. There are many solutions to this set of equations, as Vink notes. Clearly, then, one can find transition probabilities for the second extreme case (constant eigenprojections, evolving probabilities) by defining a joint current, $J_{j|i}$, satisfying the continuity equation, and choosing a solution of (4.23).

There do not seem to be generally compelling conditions that lead to a unique expression for $J_{j|i}$ and $T_{j|i}$. Therefore, I shall simply provide one example of expressions for $J_{j|i}$ and $T_{j|i}$ that satisfy the equations above.[53] I begin by choosing a simple way to satisfy the continuity and master equations, namely, adopting Bell's choice for a solution to (4.23):[54]

$$T_{j|i} := \max\left\{0, \frac{J_{j|i}}{p_i}\right\}. \tag{4.24}$$

One easily checks with (4.20) that in this case also

$$T_{i|j} = \max\left\{0, \frac{J_{i|j}}{p_j}\right\}. \tag{4.25}$$

The probability to stay in the same state, $T_{j|j}dt$, follows from normalization:

$$\sum_j T_{j|i}dt = 1. \tag{4.26}$$

Now we may begin looking for a current.

As I mentioned earlier, the existence of the maps κ_t can be used to transform away the time-dependence of $S(t)$. This fact can be put to good use in finding a 'natural' expression for the current, namely, one that generalizes the standard Schrödinger current, given by

$$J_{j|i}(t) = 2\text{Im}\left[\langle\Psi(t)|P_jHP_i|\Psi(t)\rangle\right], \tag{4.27}$$

where H is the Hamiltonian for the total system, and $|\Psi\rangle$ is the state of the total system. This current is readily seen to satisfy the continuity equation, because

$$\dot{p}_j(t) = 2\text{Im}\left[\langle\Psi(t)|P_jH|\Psi(t)\rangle\right]. \tag{4.28}$$

However, when the p_j are time-dependent, the time-derivative of $p_j(t)$ is not given by (4.28). In this case, (4.27) will not do.

However, there is a trick for generating a satisfactory current in this case from the standard current, (4.27).[55] Because the elements of $S(t)$ are mutually orthogonal, each κ_t must, in fact, correspond to some unitary operator, $O(t)$, on the Hilbert space. Using this fact, we may define

$$|\Psi'(t)\rangle := O^\dagger(t)|\Psi(t)\rangle. \tag{4.29}$$

Because $O(t)$ is unitary, it does not change the coefficients of $|\Psi\rangle$ when expanded in any basis such that every projection in $S(t)$ is spanned by some elements in the basis. In other words, $|\Psi'(t)\rangle$ differs from $|\Psi(t)\rangle$ only in the fact that its definite-valued projections are time-independent (and are, in fact, given by the definite-valued projections for $|\Psi(t)\rangle$ at time $t = 0$). The probabilities attached to these time-independent projections are the *same* as the probabilities attached to their time-dependent images under the map $O(t)$.

However, we already know how to write down a current for the time-independent case. So the obvious strategy is to write down the current for $|\Psi'(t)\rangle$, then translate the result *back* in terms of the time-dependent projections, again using the map $O(t)$. In this way, we will have derived a current for $|\Psi(t)\rangle$, which nonetheless gives rise to time-dependent definite-valued projections.

Following through with this procedure, one arrives at the result:

$$J_{ji}(t) = 2\mathrm{Im}\left[i\langle\Psi(t)|P_j(t)H(t)P_i(t)|\Psi(t)\rangle\right]$$
$$+\langle\Psi(t)|\left(\dot{P}_j(t)P_i(t) - \dot{P}_i(t)P_j(t)\right)|\Psi(t)\rangle. \qquad (4.30)$$

Therefore, (4.30) is a natural way to generalize the Schrödinger current. Note in particular that when $\dot{P}_i(t) = \dot{P}_j(t) = 0$, (4.30) reduces to the standard current, (4.27).

However, we should not get too carried away with this result — although (4.30) is certainly a natural generalization of (4.27), there is, apparently, nothing special about (4.27) in the first place — it is one among many solutions to the continuity equation.

Plenty more could be said about dynamics for modal interpretations, but I shall forego a deeper discussion here. What I have said should be enough to give the idea of how to construct a dynamics for any modal interpretation that chooses a faux-Boolean algebra of definite-valued properties at each time.

4.3 Probabilities in the modal interpretations

It is evident that the quantum logic interpretation is, in general, indeterministic. Indeed, indeterminism must enter the picture in two ways. First, the act of observing (or, coming to know, or to believe) the value that a system has for a given observable is, in general, probabilistic, in the sense that one's previous knowledge (of, for example, the statevector of the system in question) is, in general, insufficient to predict with certainty the result of the observation. However, this form of determinism is merely epistemic (though, of course, unavoidable — recall the discussion of determinism in chapter 1). What is more interesting is that there *must* be another form of indeterminism in the theory, namely, a real (non-epistemic) dynamical indeterminism: the complex of values that a system has (one for each observable) at one time cannot, in general, determine the values it will have at any later time.

Indeed, *this* way of putting the point does not even make sense in the quantum logic interpretation, for recall that one cannot attribute pairs (or, n-tuples) of values to pairs (or n-tuples) of non-commuting observables. Hence the dynamics for a given observable cannot be influenced by the values for non-commuting observables (lest we find ouselves committed to joint transition probabillities for pairs of non-commuting observables, and thereby to the meaningfulness of attributing pairs of values to pairs of non-commuting observables). Then, clearly, the dynamics for each observable must be indeterministic. For consider a system known to be in the state $(1/\sqrt{2})(|z,+\rangle + |z,-\rangle)$ (where $|z,+\rangle$ indicates spin-up in the z-drection). In that case, the system is known to have spin-up in the x-direction (because $(1/\sqrt{2})(|z,+\rangle + |z,-\rangle) = |x,+\rangle$). Now suppose that we measure σ_z (spin in the z-direction) and suppose the result is +1. Then, by the rules of the quantum logic interpretation, we must assign to the system a new state, $|z,+\rangle$. This new state, however, also determines probabilities for the possession of either spin-up or spin-down in the x-direction — it is, after all, an epistemic probability measure — and indeed it assigns each of these values the probability 1/2. Therefore, as a result of the measurement, the system must make a transition exactly 1/2 of the time from spin-up in the x-drection to spin-down in the x-direction, and clearly nothing prior to the measurement could have determined that such a transition should occur — it must be fundamentally stochastic.

Van Fraassen's Copenhagen variant of the modal interpretation is likewise necessarily dynamically indeterministic, as is the Kochen–Dieks–Healey interpretation, as the following argument (similar to the one given above) establishes. (It is obvious that, as for the quantum logic interpretation, the single-time probabilities in these interpretations are epistemic.)

Advocates of the latter interpretation sometimes speak of a measurement as 'revealing' the value that was already present, but whatever that claim means, it cannot mean that the value was present at the beginning of the experiment.[56] Consider, for example, a situation in which a system is prepared in the state $|x,+\rangle$, and then subjected to a measurment of σ_z. By the rules of the Kochen–Dieks–Healey interpretation, this system has the property 'spin-up in the x-direction' prior to measurement, but neither spin-up nor spin-down in the z-direction. Therefore the result of the measurement of σ_z cannot be determined from the start. If a system whose physical state is $|x,+\rangle$ were *guaranteed* to have spin-up (for example) in the z-direction, then the quantum-mechanical probabilities would be violated.

Finally, Bub's interpretation, which also has epistemic single-time probabilities, *may* have either deterministic or indeterministic dynamics. It has been shown that in the limit as R becomes a continuous observable, it can happen that Bub's interpretation admits deterministic as well as indeterministic dynamics.[57] The intuitive reason that the arguments above fail for Bub's interpretation is that they involved a *change* in the definite-valued observable over time: in the case of the quantum logic interpretation, this change was little more than a change in perspective (from considering σ_z to considering σ_x); in the case of the other modal interpretations, this change was a change in the set, S, used to generate a faux-Boolean algebra of properties.

However, there is another question about probabilities faced by modal interpretations. I have already broached it in the discussion of van Fraassen's modal interpretation, and will only recall it here. The question is whether the projection postulate holds in these interpretations. My contention is that if one does *not* use the quantum-mechanical state to generate a dynamics, then one *should* adopt the projection postulate, because in this case the quantum-mechanical state is nothing more than an epistemic probability measure over a system's faux-Boolean algebra of properties. If, on the other hand, the quantum-mechanical state is used to generate a dynamics, then one should take *that* role to be the primary one. The fact that the quantum-mechanical state can also be used to generate probability measures is, from this point of view, a happy consequence of the dynamics, not an independent postulate. However, note that if the dynamics is deterministic, then this consequence could never hold, because in that case, the dynamics will generate no probabilities. Exactly this problem will confront us in the next chapter, where we consider Bohm's theory, which is just a kind of modal interpretation. Indeed, it is Bub's modal interpretation with R taken to the be position observable and with a determinstic dynamics.[58]

5

The Bohm theory

In 1952 Bohm introduced a reinterpretation of non-relativistic quantum mechanics which, since then, has seen considerable development.[1] This 'causal interpretation' permits a description of quantum processes in which no reduction of the wave function occurs. Rather than being a process requiring its own postulate, measurement in Bohm's theory is described by the general dynamics. As given by Bohm in 1952, his theory's central feature is that particles follow continuous trajectories (whether observed or not), determined by a field that is always associated with the evolution of a particle. This 'ψ-field', $\psi(\mathbf{x}, t) = R(\mathbf{x}, t)e^{iS(\mathbf{x},t)}$ ($\hbar = 1$), satisfies the standard Schrödinger equation (just as the electromagnetic field satisfies Maxwell's equations) and therefore itself evolves deterministically.

5.1 Bohm's original idea

I begin by explicating Bohm's idea in terms of a two-particle wave function, and for simplicity I assume that the particles have equal mass. The Schrödinger equation in this case is:

$$i\frac{\partial \psi}{\partial t} = \left[-\frac{1}{2m}\left(\nabla_1^2 + \nabla_2^2\right) + V\right]\psi. \tag{5.1}$$

Now substitute the ψ-field into (5.1) and let $p = R^2 = |\psi|^2$. (I suppress the arguments of R and S for notational simplicity.) Straightforward calculation yields two equations corresponding to the real and complex parts of the Schrödinger equation, respectively:

$$\frac{\partial S}{\partial t} + \frac{(\nabla_1 S)^2}{2m} + \frac{(\nabla_2 S)^2}{2m} + V - \frac{1}{2m}\frac{(\nabla_1^2 + \nabla_2^2)R}{R} = 0, \tag{5.2}$$

$$\frac{\partial p}{\partial t} + \nabla_1 \cdot \left(p\nabla_1\frac{S}{m}\right) + \nabla_2 \cdot \left(p\nabla_2\frac{S}{m}\right) = 0. \tag{5.3}$$

Bohm pointed out that one can interpret (5.2) as a classical Hamilton–Jacobi equation,

$$H\left(q_1\ldots q_6; \frac{\partial S}{\partial q_1},\ldots,\frac{\partial S}{\partial q_6};t\right) + \frac{\partial S}{\partial t} = 0, \tag{5.4}$$

where q_i $(i = 1,\ldots,6)$ are the generalized coordinates. In making this interpretation, one defines a 'quantum potential'

$$U = -\frac{1}{2m}\frac{(\nabla_1^2 + \nabla_2^2)R}{R}, \tag{5.5}$$

so that (5.2) may be written

$$\frac{\partial S}{\partial t} + \frac{(\nabla_1 S)^2}{2m} + \frac{(\nabla_2 S)^2}{2m} + V + U = 0. \tag{5.6}$$

Equation (5.6) suggests that the velocity of a particle should be given by $\nabla_k S/m = \dot{\mathbf{q}}_k$, where $\mathbf{q}_k$ is the position of particle k $(= (q_1,q_2,q_3)$ for $k = 1$ in this two-particle case). (Keep in mind that S is a function of $q_1,\ldots,q_6$, so that for each k, $\dot{\mathbf{q}}_k$ is in fact a function of $q_1,\ldots,q_6$. I shall return to this point later.) Under this assumption, (5.3) becomes

$$\frac{\partial p}{\partial t} + \nabla_1 \cdot p\dot{\mathbf{q}}_1 + \nabla_2 \cdot p\dot{\mathbf{q}}_2 = 0. \tag{5.7}$$

Equation (5.7) is in the form of a conservation equation, and Bohm accordingly interprets $p(\mathbf{q}_1,\mathbf{q}_2)$ as a probability density (in configuration space). Because (5.7) guarantees that the flow of probability is conserved, one can think of p as giving the distribution of particles in an ensemble whose motion is governed by (5.6) via the prescription $\nabla_k S(\mathbf{q}_1,\mathbf{q}_2)/m = \dot{\mathbf{q}}_k$.

The 'quantum Hamilton–Jacobi equation' can therefore be seen as determining an ensemble of possible trajectories for particles with initial momenta $\mathbf{p}_k = \nabla_k S(\mathbf{q}_1,\mathbf{q}_2)$, moving under the influence of the total potential $V + U$. The 'guiding' or 'pilot' ψ-field determines (through R) the probability density p and the quantum potential U, and (through S) the momenta of the particles.

5.2 Bohmian mechanics

Bohm's 1952 theory is important for many reasons, not the least being that it disproves the old dogma that quantum mechanics is irreducibly probabilistic. However, it suffers, in some people's eyes, because it appears to be an attempt to force quantum mechanics back into a classical-looking mold, by postulating a mysterious 'quantum potential' with bizarre properties. Note, for example, that the quantum potential does not die off with distance.

Note also that it has no source. Why postulate such an entity just to make quantum mechanics look classical?

These questions are perhaps unfair — they ask for a justification for the existence and nature of the fundamental entities of a theory, and surely no theory can give an adequate answer to such questions. Nonetheless, they might raise one's suspicions about Bohm's theory, and it is useful to see that Bohm's central idea (of a deterministic theory of particles in motion) can be formulated without reference to the quantum potential.[2] I outline this formulation of Bohm's theory in this section.

5.2.1 The Bohmian equations of motion

Dürr, Goldstein, and Zanghi[3] begin with an equation like (5.1) (though for N particles) and the central Bohmian idea: the wave function governs the evolution of the positions of the particles. From here, they hope to motivate a formulation of Bohm's theory that they call 'Bohmian mechanics'.

They assume, first of all, that the wave function governs the *velocities* of the particles, and write

$$\dot{\mathbf{q}}_k = \mathbf{v}_k^\psi(q_1, \ldots, q_N), \tag{5.8}$$

where, as before, $\mathbf{q}_k$ is the position of the kth particle, and $\mathbf{v}_k^\psi$ is the velocity of the kth particle as given by the wave function, ψ. To find an expression for $\mathbf{v}_k^\psi$, they make several restrictions on $\mathbf{v}_k^\psi$, including Galilean, time-reversal, and rotational invariance. These restrictions lead them to

$$\mathbf{v}_k^\psi = \frac{1}{m_k}\text{Im}\left(\frac{\nabla_k\psi}{\psi}\right). \tag{5.9}$$

Their argument for (5.9) is not a mathematical *derivation*.[4] It is, rather, a plausibility argument, based on some physically intuitive restrictions on v_k^ψ. The hope here is not to derive Bohmian mechanics, but to show how it can be motivated in a way that perhaps the postulation of a 'quantum potential' cannot. In any case, natural or not, the Schrödinger equation plus (5.9) defines Bohmian mechanics. The latter is usually called the 'guidance condition'.

5.2.2 Interpretation of the Bohmian equations

Bohmian mechanics is a complete physical theory, in the same sense that Newtonian mechanics is. In the case of Bohmian mechanics, the complete physical state of a system is given by the positions of all the particles in the system (or, more conveniently, the position of the system in configuration space). You tell the theory the state of a (closed) system at one time, and the theory will tell you its state at any later time. Bohmian mechanics is thus a theory about *where things are*. One of the fundamental presuppositions of Bohm's theory is that the observable properties of any system supervene on position. Bohm's theory is thus highly reductionistic.

But what about mass, charge, and the like? Indeed, mass already appears in (5.9). Why is it not a fundamental physical property in Bohm's theory? Good question. The best answer I know is from an article by Brown and Anandan.[5] They consider an experiment designed to test the role of gravity in a neutron interferometry experiment. In an interferometer, there are two paths for the neutron to follow. In the experiment that they consider, the neutron is in a superposition of taking one or the other path, so that, according to orthodoxy, the neutron cannot be said to take one or the other path. (Its value for the 'which path?' observable is indeterminate.) In Bohm's theory, however, the neutron really does select one or the other path.

The catch is that the gravitational field (of the earth, for example) acts on both paths, and seems therefore to act on both 'parts' of the wavefunction. However, if the gravitational field affected *just* the particle, then it would presumably have no effect on the part of the wave function where the particle is not. They conclude that mass is not, in fact, a property of the *particle*, but is instead, in some sense, a property of the wave function.

The argument relies, ultimately, on the interpretation of the experiment in question, and I shall not enter into a discussion on that point here. Rather, I wish mainly to emphasize two points. First, this example makes it clear that to say that Bohm's theory is a theory of particles in motion is not to put an end to ontological questions. Second, based on Brown and Anandan's argument, the most plausible interpretation of Bohm's theory is that particles have *just* position, letting all other quantities be treated as parameters in the equations, rather than as properties of the particles. Brown and Anandan suggest that they are properties 'of the wave function', but, of course, the interpretational status of these parameters is yet another issue.

Bohm's theory does come in less reductionistic forms. First, there are present-day advocates of Bohm's original idea, which grants physical reality to the quantum potential. Second, there are attempts to attribute *more* than just position to the Bohmian particles. In particular, Holland advocates using the Bohmian method for observables other than position, notably, for spin.[6]

One consequence of such non-minimalism is that these authors must give some account of the quantum potential. Bohm and Hiley, in particular, are concerned to provide some understanding of the quantum potential, for, as they note, it is apparently unlike any classical potential:

we are not actually reducing quantum mechanics ... to an explanation in terms of classical ideas. For the quantum potential has a number of strikingly new features which do not cohere with what is generally accepted as the essential structure of classical physics.[7]

In several places throughout their book, Bohm and Hiley try to give us some understanding of these new features.

Mathematically, the new features are clear enough. Perhaps the most important is that the quantum potential need not fall off with distance. Indeed, multiplying R by a constant does not change the quantum potential at all, which means that 'the effect of the quantum potential is independent of the strength ... of the quantum field but depends only on its form'.[8] Bohm and Hiley's gloss on this new feature is that the quantum potential represents 'active information'; that is, it does not act as a generator of forces, but rather acts so as to inform particles how to move. They write:

Such behavior would seem strange from the point of view of classical physics. Yet it is fairly common at the level of ordinary experience. For example we may consider a ship on automatic pilot being guided by radio waves. Here, too, the effect of the radio waves is independent of their intensity and depends only on their form. The essential point is that the ship is moving with its own energy, and that the form of the radio waves is taken up to direct the much greater energy of the ship. We may therefore propose that an electron too moves under its own energy, and that the form of the quantum wave directs the energy of the electron.[9]

But how can particles be like ships? Bohm and Hiley do not shy away from the obvious answer. They suggest that perhaps 'an electron or any elementary particle has a complex and subtle inner structure (e.g., perhaps even comparable to that of a radio)'.[10]

Of course, it is far from clear how this suggestion does 'not cohere with what is generally accepted as the essential structure of classical physics'. The example that Bohm and Hiley use — and they certainly appear to take it seriously — is purely classical. However, more important here is whether any of this talk helps us to understand the guidance condition.

Apparently it does not. The idea that particles are capable of 'self-motion' guided by the quantum potential seems to be motivated by Newton's first law — motion that is not in a straight line requires some explanation; it cannot be 'free' motion. Classically free Bohmian particles do not, in general, move in a straight line, and so, or so it seems, the deviation needs an explanation. However, we have already learned to replace 'in a straight line' with 'along a geodesic' in general relativity. Why not, in Bohm's theory, suppose that the trajectories delivered by the guidance condition for a classically free particle have a status something like geodesics? In general relativity too, it is apparently possible to maintain a Newtonian space-time and introduce forces to account for the gravitational effects, but the strategy of modifying the structure of space-time instead seems at least as explanatory. Similarly, why not drop (or modify) Newton's laws here, rather than remain within an essentially Newtonian framework and introduce new forces? In any case, doing so seems at least as explanatory as presupposing Newton's first law and explaining the deviation from a 'straight line' in terms of 'self-motion'.

Holland would object that without the quantum potential, it is impossible to understand the classical limit, which, he says is given by the criterion that the quantum potential, Q, be small with respect to the classical potential, V (with a further proviso — see below).[11] However, the classical limit can just as well be given by the criteria (mathematically equivalent) that the 'geodesics' of Bohm's theory approach the classical Newtonian straight lines. Again there does not seem to be any loss of understanding (though in both cases, it is far from clear that an explanation of the classical limit has been provided at all). Indeed, given that the strength of the quantum potential is less important than its form, the classical limit must more precisely be given by $Q \ll V$ and $\nabla Q \ll \nabla V$ (as Holland recognizes[12]). Then whatever feeling of understanding we thought we got from the idea that classical behavior emerges whenever the quantum potential is 'small' is lost.

Similarly, as I mentioned earlier, Holland rejects the 'minimalist' idea (accepted by Bohm and Hiley) that the only fundamental property possessed by a Bohmian particle is position. Holland instead wishes to endow the particle with a host of properties, the account for each of which is similar to the account of its position — namely, that the particle has one of a number of possible values for, say, angular momentum, with the probability for having any given value being given by its wave function. Here too, Holland tries to convince us that the minimalist approach is unsatisfactory. He suggests again that the move to the classical limit is unintelligible if we do not endow the particles with the full set of classical properties from the start. For example, he derives a Hamilton–Jacobi equation from the Pauli equation (just as was done above for the Schrödinger equation), then shows that when the quantum potential in this equation is negligible, the equation describes an ensemble of classical dipoles. He asks: 'If this [i.e., the Pauli equation with $Q \ll V$ (and $\nabla Q \ll \nabla V$?)] indeed represents the classical limit, why should the particle's intrinsic angular momentum cease to be well defined when one passes to the quantum domain?'[13]

The question is aimed at the minimalist approach of Bohm and Hiley (and Bell), but misses its mark. Although it is crucial for any interpretation of quantum mechanics to recover the phenomenon of classicality, it is not incumbent upon any interpretation to recover the classical ontology. In particular, if the minimalist approach can recover the motions of classical objects, then there is no need for it to suppose further that those motions are accompanied by the full panoply of classical properties.

Hence the minimalist–reductionist view seems at least as good as the interpretations advocated by Bohm and Hiley, and by Holland, while avoiding the need to explain the bizarre properties of the quantum potential, and avoiding the suspicion that Bohm's theory is motivated by too strong a desire to hold on to Newton's equations. Of course, as I said before, even the minimalist version of Bohm's theory raises significant interpretational — especially ontological — questions. However, these questions I must leave aside.

5.2.3 Bohmian mechanics and quantum probability

In any case, whether one adopts the minimalist version or not, Bohm's theory is completely deterministic. How do probabilities enter the theory? *In principle* they need not. If we knew the wave function of a system, and its initial position, then we could use Bohm's theory to predict exactly where the system would be (in configuration space) at a later time. Of course, in practice we do not know a system's initial position exactly, and therefore, we must assign a probability distribution over possible initial positions. The hypothesis made by Bohm's theory is that the intial distribution is always given by $|\psi(q_1,...,q_N)|^2$. If that hypothesis is true, then because of the usual quantum-mechanical continuity equation (derived from the Schrödinger equation), the distribution will remain $|\psi(q_1,...,q_N)|^2$.

However, although Bohm's theory includes the usual quantum-mechanical rule for making probabilistic predictions, the probability theory one gets from Bohm's theory differs from quantum probability in one respect: like modal interpretations, Bohm's theory does not take the algebra of events to be $L_{\mathscr{H}}$. Instead, because it is a theory of particles in motion, the events are the Borel rectangles in $3N$-space. Apart from this modification, however, quantum probability is taken over entirely into Bohm's theory. (Hence the probability of an event in Bohm's theory for a system in the state W is still given by p^W.)

To see that quantum probability restricted to the Bohmian algebra of events becomes classical, it is helpful to look again at conditional probabilities, and see that they do indeed behave classically. If the quantum-mechanical state of the system at time 0 is $|\psi\rangle\langle\psi|(= P_\psi)$, then the quantum probability that the system is in the Borel rectangle E', given that it is in E, is

$$\frac{\text{Tr}[P_E^R P_\psi P_E^R P_{E'}^R]}{\text{Tr}[P_\psi P_E^R]},\qquad(5.10)$$

where R is the position operator and P_E^R is the projection onto the span of all P_x^R for $x \in E$. However, restricted to just the P_E^R (for every Borel rectangle, E), (5.10) is a classical conditional measure, because P_E^R and $P_{E'}^R$ belong to a Boolean algebra, so that their product is just $P_{E \cap E'}^R$. Hence (5.10) may be written:

$$\frac{p^\psi(E \cap E')}{p^\psi(E)},\qquad(5.11)$$

which is, of course, just the classical conditional probability of E given E'.

What about transition probabilities? In the standard theory, they are calculated with the help of the projection postulate. Bohm's theory too can calculate transition probabilities this way, despite the fact that states *never* really collapse in Bohm's theory. To see why, keep in mind that Bohm's theory adopts the usual quantum-probabilistic definition for conditional probabilities. Now, suppose we are told that the system was in E at time 0, and we were asked for the probability that it will be in E' at time t. Given a state, $|\psi\rangle$, $p^\psi(\cdot)$ is a probability distribution over the algebra of Borel sets in configuration space. The support of this distribution tells us where the system *might have been* at time 0. Adding to this our information that E occurred, the 'obvious' thing to do is to restrict the distribution p^ψ to E, renormalize, and use the resulting distribution to calculate the probability for E' at time t. This procedure is exactly what the projection postulate says to do.

However, this procedure works only for a subtle reason. Note that in general we may *not* simply project out part of W, and use it to calculate how a system will evolve. The reason is that, in Bohm's theory, $|\psi\rangle$ never collapses — it is always playing its (primary) role of determining v_k^ψ. However, in typical situations where we are in a position to learn the location of a system, macroscopically many particles are involved, and just as in standard quantum mechanics, a collapsed state will yield different predictions from an uncollapsed superposition only if we can bring two of the terms in the superposition back together in configuration space. When macroscopically many particles are involved, each term in the superposition is specified by a great many degrees of freedom (three for each particle), and then recombining of the waves is not practically possible. Therefore, in practice, we may use the projection postulate. Bohm's theory thus *explains* the success of the projection postulate, without falling prey to the objections that beset it (because, in Bohm's theory, the projection postulate is not taken as fundamental — it is a consequence of the theory together with the fact that observation typically involves macroscopically many particles). It also provides a criterion for when to apply the rule, namely, when a Bohmian event, E, has occurred in circumstances where interference effects can be ruled out on practical grounds.

5.3 Classical experience in Bohmian mechanics

There are a number of questions that the account thus far has left unanswered. I have mentioned a few. The discussion of the previous section raises (at least) one more, namely, whether *all* experience involves macroscopically many particles, and if not (as seems likely[14]), whether Bohm's theory can adequately account for our experience of microscopic or mesoscopic bodies. I shall not address this question here (apparently an answer similar to the one given for CSL could be given) and, indeed, I shall raise just two further questions about Bohm's theory, only to give examples of the *sort* of discussion that I take to be important for understanding Bohm's theory. The first question is whether certain central features of classical mechanics can be recovered by Bohm's theory, and the second is whether Bohm's theory adequately accounts for quantum-mechanical chances.

I have said how Bohm's theory proposes to avoid the difficulties faced by standard quantum mechanics: it changes the algebra of events from $L_{\mathcal{H}}$ to the algebra of Borel rectangles in configuration space; it denies the eigenspace–eigenvalue link; and it provides a relatively clear criterion for when the projection postulate is effectively applicable. However, are these modifications enough to solve the really difficult problem of describing classical experience?

Because Bohm's theory is a full-fledged theory, it will be useful to discuss this question in some detail. To do so, I begin by highlighting two distinguishing features of classical systems (the following list is not comprehensive, but representative): (1) classical systems are deterministic; and (2) the structure of classical events is that of a Boolean algebra. Neither of these features is shared by standard quantum mechanics. Therefore, the problem of describing our classical experience in Bohm's theory involves in part the problem of explaining how classical systems, which in principle should be described by Bohmian mechanics, have these features.

The question is harder than it might appear. My strategy in each case will be first to show why it is not immediately obvious that Bohm's theory can recover these features of classical systems. Then I will argue that, nonetheless, it can.

5.3.1 The problem of recovering classicality

In general, in Bohm's theory, one does not know anything about the location of a system beyond what is contained in its wave function. Moreover, as Dürr, Goldstein, and Zanghì have argued (and Bohm apparently already realized in 1952),[15] measuring the present location of a system destroys the possibility of predicting its future location with any more precision than the wavefunction affords. Hence, although Bohm's theory is fundamentally deterministic, it is predictively just as indeterministic as standard quantnum mechanics.

It is essentially this predictive indeterminism that represents Bohm's way of satisfying Heisenberg's uncertainty relations. Bohm's account of the uncertainty relations is therefore quite classical: the act of measuring position disturbs the velocity of a particle in an uncontrollable way (which is why one cannot use information about the present location of a system to make accurate predictions about its future positions). The non-classical element here is just that the disturbance is *unavoidable*. The uncertainty relations are a theorem of Bohm's theory. Hence the predictive indeterminism of Bohm's theory is fundamental and unavoidable. How does this fundamental predictive indeterminism become predictive determinism for classical events? In the next subsection I shall address this question.

What about the Boolean structure of classical events? As we have seen, the fundamental physical events in Bohm's theory are (represented by) the Borel rectangles in configuration space, which form a Boolean algebra; but even so, whether Bohm's theory can recover the Boolean structure of classical events is not a trivial question. To see why, recall that $L_{\mathcal{H}}$, the algebra of events in standard quantum mechanics, is non-distributive, and for that reason (and only that reason) non-Boolean. How does Bohm's theory reproduce the non-Boolean structure of quantum events?

The answer begins with the following fact.[16] Given a set of projections $\{P, P_i\}$, if $P \wedge (\bigvee_i P_i) \neq \bigvee_i (P \wedge P_i)$ then P fails to commute with at least one of the P_i. Therefore, non-distributivity can be 'blamed' on non-commutivity. Furthermore, as I illustrate in an example below, Bohm's theory does *not* allow simultaneous assignment of values for non-commuting observables, and therefore does not violate the non-distributivity of quantum events.

The reason is that Bohm's theory is contextual: the value of some observables for a system depends, in part, on the environment of the system. This point is easiest to see in an example, for which it is convenient to use the following formulation of Bohm's theory, equivalent to the one I have given above. Let $\mathbf{q}$ be a variable that takes as values points in configuration space. Let $\rho(\mathbf{q}, t) = \Psi^*(\mathbf{q}, t)\Psi(\mathbf{q}, t)$ and let $j(\mathbf{q}, t)$ be the probability current derived from the Schrödinger equation in the usual way. Then define

$$v^\psi(t) = \frac{j(\mathbf{q}, t)}{\rho(\mathbf{q}, t)}, \tag{5.12}$$

where v^ψ is the velocity of the system in configuration space.

Now, consider a spin-1/2 particle and the quantum events $P_{z,+}$, $P_{x,+}$, and $P_{x,-}$, where $P_{u,\pm}$ represents spin-± 1 in the u-direction. These projections constitute a non-distributive set: $[P_{z,+} \wedge (P_{x,+} \vee P_{x,-})] \neq [(P_{z,+} \wedge P_{x,+}) \vee (P_{z,+} \wedge P_{x,-})]$. Keep in mind that because Bohmian particles (in the minimalist interpretation that I prefer) do not have intrinsic spin, to say that a particle 'has' spin-up in the z-direction just means that the particle is headed for (or, has arrived at) the detector set to 'indicate' spin-up in the z-direction.

How does such a measurement occur in Bohm's theory? Bell has exhibited the current for measurements of spin in the u-direction:[17]

$$j(\mathbf{q}, t) = \psi^\dagger(\mathbf{q}, t)g(t)\sigma_u\psi(\mathbf{q}, t), \tag{5.13}$$

where $g(t)$ characterizes the spin–orbit coupling and σ_u is the spin-operator in the u-direction. From (5.12) and (5.13), it is clear that the velocity field depends on whether σ_z or σ_x is measured. Hence, if σ_z is to be measured, one cannot assign a value to σ_x. (A particle cannot be headed for a detector that is not there!) This contextuality in Bohm's theory allows it to recreate quantum non-distributivity, in the sense that Bohm's theory refuses to define simultaneous values for all events in a non-distributive set. (It is also how Bohm's theory avoids the Kochen–Specker theorem — in Bohm's theory, not all of $L_\mathcal{H}$ is definite-valued at any given time.)

5.3.2 Recovering classicality

To see whether Bohm's theory can recover the features of classical systems, it will help to try to characterize what is meant by 'the events of our classical experience'. The objects that we experience typically have large numbers of particles and are well localized in space. Hence we may suppose that classical events are those where many particles (perhaps $\gg 10^{23}$) are inside a macroscopic region. Such events are indeed the events of our everyday experience — the chair is here; the billiard ball is there. (As I mentioned before, they might not be the *only* events of our everyday experience, but they are clearly typical, and I shall not consider other types of event here.[18]) How is it that, when we restrict Bohm's theory to such events, it exhibits the classical features described above?

Before I address this question, I emphasize that I do not aim to *derive* a classical limit, in the sense of showing that there *is* a classical regime. My aim is to show how classical features emerge from a Bohmian description of macroscopic objects. However, it is worth noting that there are arguments to suggest that in a Bohmian universe there will appear a domain in which macroscopic events occur. Specific examples can be found in the literature, but Bohm's theory can also rely on similar results from studies of decoherence in quantum mechanics.[19]

In any case, however we get to a domain of macroscopic events, it is by no means obvious that such events, when considered within Bohm's theory, will exhibit classical features. Below, I suggest that in Bohm's theory, classicality (or at least those features of classicality that I have considered, namely, predictive determinism and a Boolean structure of events) does emerge.

Regarding predictive determinism, recall that Bohm's theory, although fundamentally deterministic, is in principle predictively indeterministic at the level of (most) quantum events. What needs arguing now is that Bohm's theory is once again deterministic (in practice — we know that it must be in principle) at the level of classical events. There are really two questions at issue. First, can Bohm's theory underwrite the phenomenological success of classical physics? What I have in mind here is the following procedure: translate a classical state-description, an initial state, into Bohm's theory; calculate the evolution of the system; translate the final state back into the language of classical physics. The question is whether we get a deterministic relationship between initial and final *classical* states. Second, ignoring the *practice* of classical physics, do large systems evolve deterministically?

To answer the first question, suppose that we have specified a classical event and a momentum, or, more realistically, a narrow range of possible momenta. How does one treat such information in Bohm's theory? The occurrence of a classical event allows one to write down a wave function, $\psi(\mathbf{q}_1, \ldots, \mathbf{q}_N)$, with support just over the region of $3N$-space named in the event. Furthermore, because further information about the location of the system is unavailable, one may take as most reasonable a uniform distribution of ψ across its support. Indeed, probably the best way to represent the state of a system localized in a region, R, is with the projection operator representing the event of being localized in R (suitably normalized).

This event, taken as the state of the system, generates a uniform distribution over R. This uniform distribution, together with the specification of a precise, or nearly precise, momentum, entails that the velocity field for the system is roughly uniform over the support of ψ, and that uniformity is enough to regain determinism. When the velocity field is roughly uniform, our ignorance of the system's precise location inside the wavepacket is irrelevant to the prediction of future classical events. Moreover, because Bohm's theory is deterministic at the classical level, it follows that any probabilities at that level are purely epistemic.

That argument looks good, but it is perhaps more contentious than it looks, for it relies on the fact that in Bohm's theory the wave function plays two conceptually distinct roles — the (epistemic) role of providing a probability measure, and the (ontological) role of determining the velocity field for a system. In general, one might question whether information about the *epistemic* probability distribution for a system implies anything about its *actual* wave function. On the other hand, there does not appear to be much room in Bohm's theory to *deny* that implication.

The second question about determinism is easily answered, using the fact (discussed earlier) that the projection postulate is *effectively* applicable to large systems in Bohm's theory. Consider a system in the state

$$|\Psi\rangle = \sum_i c_i |\varphi_i\rangle, \tag{5.14}$$

where the $|\varphi_i\rangle$ (more precisely, the projections onto the subspaces spanned by the $|\varphi_i\rangle$) represent macroscopically distinct classical events. The evolution of a large system in the superposition (5.14) is *effectively* the evolution (as given by Schrödinger's equation) of a *single* element in the superposition, because the various $|\varphi_i\rangle$ will not interfere. (That is, we can consider the system to lie in just one of the $|\varphi_i\rangle$.) However, that evolution is evidently deterministic. Indeed, by Ehrenfest's theorem, it is exactly the classical evolution.[20] Hence the predictive indeterminism of Bohm's theory disappears at the classical level.

Contextuality, and therefore non-commutativity, and therefore non-distributivity, and therefore non-Booleanism, also disappear at the classical level. That contextuality disappears for classical events is a trivial consequence of the fact that all classical events commute. Of course, this fact — i.e., that commutivity implies Booleanism — holds in standard quantum mechanics as well, but the point here is that in Bohm's theory, there is a physical explanation to accompany the mathematics. Contextuality disappears when one can simultaneously assign values to all observables, and all classical observables (or better, classical events, as defined here) do commute with one another. Moreover, there is a physically intuitive explanation for why commuting events, but not non-commuting events, can simultaneously be assigned values in Bohm's theory: the different apparatuses used to measure each member of a non-commuting set affect the velocity field of the measured system differently. In standard quantum mechanics, we are told (and it is proven to us mathematically) that there is no joint probability function for non-commuting events, but rarely if ever are we given a physical explanation for why these quantities are not defined.

The discussion here has been brief and rapid, but it should be sufficient to make it clear both that the question of whether Bohm's theory recovers the features of the classical world is not a trivial one, and that, in the end, Bohm's theory *does* appear to recover those features.

5.4 Probability in Bohm's theory

It does not follow that Bohm's theory is completely satisfactory. Thus far, I have largely ignored the most difficult question facing Bohm's theory, namely: Why is $|\psi|^2$ a probability?

There are two questions here. First, why should probability be given by $|\psi|^2$ rather than anything else? Second, why should ψ, which has the primary role of determining the velocities of particles, have anything to do with probability? We may also wonder both why there are probabilities at all in Bohm's theory, and what they mean.

Work by Dürr, Goldstein, and Zanghi and by Valentini makes some headway on these questions.[21] Dürr *et al.* begin by noting that, in Bohmian mechanics, the only system that always has a wave function is the universe. Call the wave function of the universe '$\Psi(\mathbf{q})$', where $\mathbf{q}$ is a variable in the configuration space for the universe. When do subsystems (to which we typically attribute wave functions) really have a wave function? Dürr *et al.* make the following definition, letting the system in question be represented by the variable, $\mathbf{x}$, in its configuration space and the rest of the universe be represented by the variable $\mathbf{y}$ in the configuration space for the rest of the universe:

Effective wave function: If

$$\Psi(\mathbf{q}) = \Psi(\mathbf{x}, \mathbf{y}) = \psi(\mathbf{x})\Phi(\mathbf{y}) + \Psi^{\perp}(\mathbf{x}, \mathbf{y}) \tag{5.15}$$

where the $\mathbf{y}$-supports of $\Phi(\mathbf{y})$ and $\Psi^{\perp}(\mathbf{x}, \mathbf{y})$ are macroscopically distinct, *and* if the actual value of $\mathbf{y}$ (which, of course, always exists in Bohm's theory) lies in the support of $\Phi(\mathbf{y})$, then the system described by $\mathbf{x}$ has *effective wave function* $\psi(\mathbf{x})$.

The motivation for this definition is clear. If the **y**-supports of $\Phi(\mathbf{y})$ and $\Psi^{\perp}(\mathbf{x}, \mathbf{y})$ are macroscopically distinct, then it will at least take some time to bring them back together, so that, meanwhile, no interference effects will occur, and the system described by **x** will be effectively governed by $\psi(\mathbf{x})$. A typical situation is the end of a measurement, where **x** describes the measured system, and **y** the apparatus, which has, at the end of the measurement, a definite value in the support of $\Phi(\mathbf{y})$.

Dürr *et al.* then consider an ensemble of systems, each in the state $\Psi(\mathbf{q})$, whose configurations are distributed according to $|\Psi(\mathbf{q})|^2$. They then consider a subensemble, selected as all systems whose **y**-support is in $\Phi(\mathbf{y})$, and for which, therefore, the **x**-subsystem has the effective wave function $\psi(\mathbf{x})$. Under these conditions, they prove[22] that the probability that $x \in dx$ is $|\psi(\mathbf{x})|^2 dx$.

This result (apart from how it is used) is important in itself. It shows that the standard practice in quantum mechanics is justifiable in Bohm's theory, in the sense that if probabilities are given by $|\Psi|^2$ for a collection of systems, then the same is true for a collection of subsystems selected from them, for example by a measurement or preparation.

Dürr *et al.* try to use this result in the service of the Bohmian postulate that whenever a subsystem of the universe has effective wave function $\psi(\mathbf{x})$, the probability that $\mathbf{x} \in dx$ is given by $|\psi(\mathbf{x})|^2 dx$. Their argument relies on what I shall call the 'hypothesis of the initial condition':

Hypothesis of the initial condition: Given the set $Q = \{\mathbf{q}^{(i)}, \mathbf{q}^{(ii)}, \ldots\}$ of all possible initial distributions of the universe, the probability measure over this set is given by $|\Psi_0(\mathbf{q})|^2$ ($\mathbf{q} \in Q$), where $\Psi_0(\mathbf{q})$ is the wave function of the universe at the initial time.

In this section, I will refer to this hypothesis as just 'the hypothesis'. Note that because $\Psi(\mathbf{q})$ satisfies the continuity equation, if the distribution at the initial time is given by $|\Psi_0(\mathbf{q})|^2$, then at later times, t, it is given by $|\Psi_t(\mathbf{q})|^2$.

The result of Dürr *et al.* shows that, given the hypothesis, the usual quantum probabilities follow. That is, given the hypothesis, if the system described by **x** has effective wave function $\psi(\mathbf{x})$, then the probability that $\mathbf{x} \in dx$ given that **y** is in the support of $\Phi(\mathbf{y})$ is $|\psi(\mathbf{x})|^2 dx$.

In fact, their result shows more: namely, that if the system described by **x** has effective wave function $\psi(\mathbf{x})$, then the probability that $\mathbf{x} \in dx$ given the actual value of **y** (a specific point somewhere in the support of $\Phi(\mathbf{y})$) is $|\psi(\mathbf{x})|^2 dx$. It follows that epistemic probabilities in Bohm's theory are necessarily non-trivial ('necessary' in the physical sense). The reason is that the value of **y** could, in principle, contain complete information about the entire universe, excluding the system described by **x**. Hence, assuming (as Dürr *et al.* do) that knowledge is ultimately grounded in the physical configuration of the universe, and that knowledge *about* the system described by **x** is grounded in the configuration of the rest of the universe, then — no matter what this knowledge (configuration) might be — the probability that **x** ∈ *d***x** conditioned on this knowledge is still $|\psi(\mathbf{x})|^2 d\mathbf{x}$. I have already mentioned and discussed this result in section 5.3.1.

Therefore, Dürr *et al.* have shown that *given* the hypothesis, a universe governed by Bohmian mechanics will necessarily obey the usual quantum-mechanical probabilities.

However, this argument relies on the hypothesis, and there is plenty of room to wonder not only whether the hypothesis is true, but what exactly it means. What could be the meaning of a probability distribution over possible initial configurations of the universe? Is it an *actual* distribution? If so, then we must be prepared to countenance an actual multiplicity of universes. Is it an epistemic distribution? Perhaps, but then we are left with the very difficult question of why $|\Psi_0\rangle$ should have anything to do with our epistemic probability distribution over possible initial configurations.

Dürr *et al.* do have a name for what the 'probability' of an initial condition for the universe means — they call it 'typicality'. However, that name only pushes the question back one step. Normally when we say that something is 'typical' we mean that it is common, or that it occurs frequently, or habitually. None of these meanings applies to the universe, which is unique, and, of course, to say that 'typical' means 'probable' is circular.

Within the general approach of beginning with the universe, and working down to the statistics for subsystems, I see only three ways to interpret the hypothesis. First, one could countenance a true multiplicity of universes, distributed according to $|\Psi(\mathbf{q})|^2$. I shall not consider this possibility any further. Second, $|\Psi(\mathbf{q})|^2$ could be the 'objective chance' that the universe began in the distribution *q*. Insofar as one could make sense of an objective chance for some initial distribution of the universe — and I shall not consider here whether one can do so — this interpretation is acceptable, though the hypothesis must then be considered a postulate, probably without the possibility of justification. Third, $|\Psi(\mathbf{q})|^2$ could be a subjective probability distribution over the set, $Q = \{\mathbf{q}^{(i)}, \mathbf{q}^{(ii)}, \ldots\}$ of all possible initial distributions of the universe. In this case, the hypothesis would be not so much a postulate as a suggestion — if you make your subjective probability $|\Psi(\mathbf{q})|^2$ then you will make good statistical predictions about the states of subsystems of the universe.

However, apart from the fact that the hypothesis guarantees empirical adequacy for Bohm's theory, why should it be believed? What justification can Bohm's theory provide for it?

Two very different sorts of answer have been suggested. Dürr *et al.* have argued that the quantum-mechanical distribution, $|\Psi(q)|^2$ is the only natural measure over initial conditions, because it is the only 'equivariant' measure.[23] An equivariant measure is just a measure that obeys the continuity equation, i.e., one that is preserved by the Bohmian equations of motion.

There are two problems with this argument. First, Dürr *et al.* have not shown that $|\Psi(q)|^2$ is the *only* equivariant measure — the existence of another equivariant measure (and especially, one that is a function of Ψ) would clearly undermine their argument. As far as I know, whether $|\Psi(q)|^2$ is the only equivariant measure is an unanswered question.

Second, and more serious, it is not at all obvious why equivariance is a preferred property of measures over the possible initial distributions. Equivariance is a *dynamical* property of a measure, whereas the question 'Which initial distribution is the correct one?' involves no dynamics, nor is it clear why dynamical properties of a measure are relevant.

Of course, apart from these specific problems, it remains unclear what the criteria for an initial measure should be. Imagine that you are creating the universe. You have the set Q of initial distributions from which to choose. Suppose you decide to choose randomly, according to some probability distribution over Q. Which probability distribution do you choose? What criteria do you use to choose a probability distribution? I cannot imagine that there is any uniquely reasonable criterion sufficient to specify just one distribution over Q.

Valentini has suggested another answer to the problem of justifying the hypothesis, by proving a quantum-mechanical version of the H-theorem.[24] Consider an ensemble of systems, all with the quantum-mechanical state $\Psi_0(q)$, and distributed according to the probability distribution, $\pi(q)$, which is not necessarily the quantum-mechanical distribution. Valentini defines a 'subquantum entropy' as:

$$S = -k \int d^{3N}q \, \pi(q) \ln \left(\pi(q)/|\Psi(q)|^2 \right), \qquad (5.16)$$

where k is Boltzman's constant. Now, given a coarse graining of π and Ψ, denoted '$\bar{\pi}$' and '$\bar{\Psi}$', Valentini shows that the coarse-grained entropy,

$$\bar{S} = -k \int d^{3N}q \, \bar{\pi}(q) \ln \left(\bar{\pi}(q)/|\bar{\Psi}(q)|^2 \right), \qquad (5.17)$$

increases with time. Assuming that it reaches its maximum value (which turns out to be 0), we find that $\bar{\pi} = |\bar{\Psi}|^2$.

The main idea, then, is that we need not assume the hypothesis after all. We can allow the initial distribution, π, to be anything. Eventually, it will become (very close to) the quantum-mechanical distribution. The argument is therefore very similar to 'mixing' arguments in classical statistical mechanics. Why are the molecules in the room distributed according to the classical equilibrium distribution? Well, imagine that they were initially all in the corner of the room. Eventually, they would become distributed roughly evenly throughout the room.

Valentini's argument is therefore perfectly parallel to the corresponding arguments in classical statistical mechanics. Hence, although his result is important, we may wonder how far it goes towards clarifying and justifying the use of the quantum-mechanical probability distribution within Bohm's theory. There are (at least) three obstacles to using Valentini's argument thus. All of them are familiar from discussions of the foundations of classical statistical mechanics.

First, just as Dürr *et al.* did, Valentini must assume that $|\Psi_0(\mathbf{q})|^2$ is the only equivariant measure. If there is *another* equivariant measure, then the status of Valentini's argument is unclear at best. Nonetheless, let us allow, for the sake of the argument, that $|\Psi_0(\mathbf{q})|^2$ is the only equivariant measure. There remain two further obstacles.

Second, why are we allowed to assume that the coarse-grained entropy has been maximized by now? There is no time scale in Valentini's argument (as there is not in its classical analogues), and therefore although the argument shows that $\bar{S}$ increases in time, we have no idea how fast it increases. Indeed, the argument is compatible with a universe in which $\bar{P} \neq |\bar{\Psi}|^2$ arbitrarily far into the future. On what basis (apart from dubious anthropic principles) may we claim that we have been lucky enough that coarse-grained entropy has, in fact, been maximized?

Third, what is the justification for the coarse-graining? It is a kind of 'uniformity' assumption. That is, it is the assumption that actual distributions inside the cells created by coarse-graining are sufficiently uniform, and are not 'too pathological' — 'too pathological' meaning here 'such as not to permit mixing'. If the initial distribution of the universe was an analogue of a classically non-mixing state — e.g., all the gas molecules in a perfectly rectangular room moving exactly parallel to the floor and to one another — then the quantum-mechanical distribution would never be reached.

In the end, we do not seem to have a justification for the hypothesis. The arguments of Dürr *et al.* and Valentini are suggestive, but far from providing a satisfactory justification for the use of $|\Psi_0(\mathbf{q})|^2$ as a probability measure in Bohm's theory. The fact that probabilities are given by $|\Psi_0(\mathbf{q})|^2$ must remain a postulate in Bohm's theory.

Of course, *asking* for a justification is perhaps to impose a rather high standard of explanatory power on Bohm's theory. The point remains that it is the most successful interpretation of quantum mechanics, providing as it does a solution to the measurement problem, and a tolerably clear picture of the quantum world.

Part two

Quantum non-locality

6

Non-locality I: Non-dynamical models of the EPR–Bohm experiment

We now have under our belts several interpretations of quantum mechanics, each requiring, or anyhow advocating, some understanding of quantum probabilities. In the next two chapters, I will consider some connections among various interpretations of quantum probabilities and non-locality. I do so in the context of the well-known EPR–Bohm experiment, though it is worth emphasizing at the start that non-locality is very likely the *rule* rather than the exception for quantum-mechanical systems. Entanglement of systems occurs not only in the confines of a laboratory, but also in the course of quite typical interactions among quantum-mechanical systems.

Nonetheless, the EPR–Bohm experiment shines a bright light on the phenomenon of non-locality, and is therefore the most useful context in which to explore the relation between probability and non-locality. In this chapter, I consider models of the EPR–Bohm experiment that deliver probabilities for the various outcomes given the initial state of the pair of particles. In the next chapter, I consider fully dynamical models, i.e., ones that provide a dynamics for the complete state of the pair of particles as well as probabilities for various outcomes based on these complete states.

6.1 The EPR–Bohm experiment

The EPR–Bohm experiment is well known, but some observations about it are important for later. A standard schematic depiction is given in figure 6.1. There, a pair of particles is emitted from the source, each headed for its respective Stern-Gerlach magnet. Each magnet is set to measure spin in some direction, and the result is that the particle is deflected either 'up' or 'down', to be detected in an 'up' or 'down' region on its detector.

This picture has led to a common distinction between 'parameter-settings' and 'outcomes', the former being the orientation of the Stern–Gerlach mag-

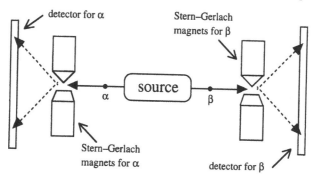

Fig. 6.1. *Standard schematic depiction of an EPR–Bohm experiment.* A pair of particles is emitted from the source, each headed for its respective Stern–Gerlach magnet.
Each magnet is set to measure spin in some direction, and the result is that the
particle is deflected either 'up' or 'down', to be detected in an 'up' or 'down' region
on its detector.

nets, and the latter being the flash at either the 'up' or 'down' region of a
detector. Hence one might be tempted to define the events in this exeriment
to be:

$M_d^n \equiv$ the event: particle n's Stern–Gerlach magnet is set in the d-direction;

$D_\pm^n \equiv$ the event: particle n's detector flashes in the spin-$\pm 1/2$ region.

We shall see in the next section that some well-known accounts of non-
locality in the EPR–Bohm experiment are couched in these terms.

However, these terms are misleading. To see why, consider the experiment
again, and consider two of the possible orientations of the Stern–Gerlach
magnet for particle α — say in the x- and y-directions. Concentrating on just
one of the particles, we see something like what is schematically depicted in
figure 6.2.

In figure 6.2 there are no single 'up' or 'down' regions of the detector.
Instead, the detector flashes *somewhere*, and to interpret this flash as 'up'
or 'down' in some direction requires knowledge of the orientation of the
magnets. Therefore, the outcomes of the experiment are better defined as:

$D_{d,\pm}^n \equiv$ the event: particle n's detector indicates spin-$\pm 1/2$ in the d-direction.

However, one can object that the superiority of the latter definition of the
outcomes is an artefact of the experiment that I have considered. True, in a
typical measurement of spin the 'up' and 'down' regions of the detector differ
for different directions of spin, but why can we not imagine a device that
merely flashes red for spin-up or green for spin-down, without indicating a
direction? In this case, the earlier definition suffices.

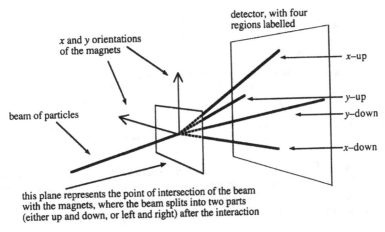

Fig. 6.2. *Schematic depiction of one wing of an EPR–Bohm experiment.* The particle approaches the magnet (represented here by the first plane) along the *z*-axis. The magnet is oriented in some direction (two of which are shown here — *x* and *y*) leading to a splitting of the particle's wave function in the *zd*-plane, where *d* is the orientation of the magnet. (Two such splittings are shown in the figure, one in the *zx*-plane, and the other in the *zy*-plane.)

This way of salvaging the earlier definition is unsatisfactory. While somebody could surely design the device in question, it is essentially a device that hides from us information that it 'knows' — i.e., information that is discernible from its complete state. Indeed, the device itself will have somehow to take account of the orientation of the magnets in order to decipher the meaning of a flash in a given region. For example, given only a flash in the 'y-up' region (as labelled in figure 6.2), it is impossible to tell whether the particle passed through magnets oriented in the y-direction, or magnets oriented anti-parallel to the y-direction, and without this information the device cannot 'know' whether to flash red or green. Therefore, the complete state of the device must somehow include the orientations of the magnets. However, then the *real* detector-events — the ones that actually occur in the device — will include the fact of how the magnets were oriented. That a device can be made to hide this information from the casual onlooker is uninteresting.

There is yet another way to salvage the old definition. Consider instead of a measurement of spin, a measurement of the polarization of photons. (See figure 6.3.) The experiment is exactly analogous to the EPR–Bohm experiment, except that there is no deflection of the beam of photons. Instead, each beam encounters a polarizer, oriented in some direction. A given photon has some chance of passing the polarizer and some chance of

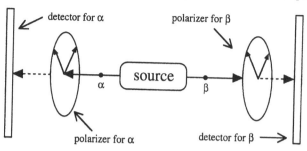

Fig. 6.3. *Schematic depiction of a polarization experiment.* The photons leave the source, and when they hit the polarizers, they either pass or not. (Each polarizer can be set to any direction, two of which are indicated by arrows on the polarizers.) The detector behind the polarizer discerns whether the photon passed.

not passing (just as a particle has some chance of having spin-up in a given direction, and some chance of having spin-down). However, whether the photon passes or not is the *only* information that is encoded in the detector behind the polarizer. Hence 'flash' or 'no flash' may be taken as complete descriptions of the detector-events for this experiment.

However, we should ask what these events mean. Again, without knowing the orientation of the polarizer, a flash at a detector is uninterpretable — it tells us *nothing whatsoever* about the state of the photon that caused the flash. Hence, in this experiment too, mere flashes are experimentally meaningless and, while the definition of the detector-events as $D_{\pm}^{n}$ is more plausible in this case, these events cannot be taken to be outcomes of any meaningful experiment — for example, measuring the polarization of a photon. Put differently, outcomes and parameters cannot be isolated from one another.

This point is, I believe, quite generic. Although 'flash' and 'no flash' are meaningful physical events, they are not meaningful outcomes for the sorts of experiment that are relevant to this discussion. The 'outcome' of an experiment already contains the 'parameter'.

6.2 Analyses of locality

6.2.1 Non-locality in standard quantum mechanics

The fact that outcomes already contain parameters might have implications for some popular analyses of 'locality' in the EPR–Bohm experiment (which is the experiment that I shall mainly consider from now on).

To see why we are motivated to consider 'locality conditions' at all in this experiment, recall that there are correlations between the results at different wings of the experiment. In the EPR–Bohm experiment, the quantum-mechanical state, $|\psi\rangle$, of the particles leaving the source is the singlet state:

$$|\psi\rangle = \frac{1}{\sqrt{2}}\left(|z,+\rangle^{\alpha}|z,+\rangle^{\beta} - |z,-\rangle^{\alpha}|z,+\rangle^{\beta}\right), \tag{6.1}$$

where $|z,+\rangle^{\alpha}$ is a vector in $\mathscr{H}^{\alpha}$ and indicates 'spin-up in the z-direction', and so on. (The direction z in (6.1) can be replaced by any direction, because the singlet state is spherically symmetric.) In this case, quantum mechanics yields joint probabilities for all orientations i and j of the two magnets — i and j are unit vectors. To simplify notation, let i_{α} be a variable ranging over ± 1 and similarly for j_{β}. Here $i_{\alpha} = 1$ is to be interpreted as 'the event $D^{\alpha}_{i,+1}$ occurs', and so on. Then for any i_{α} and j_{β},

$$p^{\psi}(i_{\alpha}, j_{\beta}) = \left|\left(\langle i, i_{\alpha}|\langle j, j_{\beta}|\right)|\psi\rangle\right|^{2}. \tag{6.2}$$

Because I shall not deal with mixed states in this chapter, I write quantum-mechanical probabilities as inner products rather than as traces.

These joint probabilities yield the much-advertised correlations between the results at different wings. Given the settings i and j, the long-run average of the product of the results at each wing will be (according to quantum mechanics) $-i \cdot j$, which is $-\cos\theta_{ij}$, where θ_{ij} is the angle between i and j. That is, for any i and j,

$$\sum_{i_{\alpha}, j_{\beta}=\pm 1} i_{\alpha}\, j_{\beta}\, p^{\psi}(i_{\alpha}, j_{\beta}) = -i \cdot j. \tag{6.3}$$

Immediately one can see that there are non-trivial correlations between the results of the measurement on the two particles. Indeed, if $\theta_{ij} = 0$ then the results are always opposite: $i_{\alpha} = -j_{\beta}$ on every run of the experiment. It is, of course, natural to ask how these correlations are produced. How do the particles 'know' enough about each other to maintain the proper correlation? The question becomes perhaps more dire when we consider the theory of special relativity, which we might take to prohibit 'influences' from one region to a space-like separated one. (For now, I shall remain deliberately vague about what counts as an 'influence', and I shall delay discussing the issue of whether relativity really does contain any such prohibition.) We might therefore be led to propose a constraint on explaining the correlation: if the measurements in different wings occur in space-like separated regions of space-time, then nothing that happens in one measurement-region of space-time 'affects' (another term of vagueness!) what goes on in the other.

How can we make this condition precise, in terms of the events defined for this experiment? I have suggested that the empirically important events are the $D^{n}_{d,\pm}$ and the empirically important quantities are the probabilities for these events. Hence the most natural locality condition appears to be that the events $D^\alpha_{i,\pm1}$ and $D^\beta_{j,\pm1}$ are probabilistically independent. That is, for any orientations, i and j, and any results i_α and j_β

$$p^\psi(i_\alpha, j_\beta) = p^\psi(i_\alpha) \times p^\psi(j_\beta). \tag{6.4}$$

Given any initial state $|\psi\rangle$ for the particles that is not a tensor product state (i.e., a tensor product of a state exclusively for particle α and a state exclusively for particle β), this condition fails for at least some i and j and some results i_α and j_β. Hence, by this definition of 'locality' at least, quantum mechanics is clearly non-local.

6.2.2 Bell-factorizability and Jarrett-factorizability

The question immediately arises of whether quantum mechanics *must* be non-local. We might hope to find some more complete state-description for the pair of particles that would screen $D^\alpha_{i,\pm}$ off from $D^\beta_{j,\pm}$ and *vice versa*, thus rendering them statistically independent. Give this complete state-description the generic name, 'λ'. Our hope, then, is that some 'hidden-variables' theory (whose hidden variables are the λ) delivers a probability distribution, denoted 'p^λ', that satisfies:

Bell-factorizability: For any λ, i, j, i_α, and j_β

$$p^\lambda(i_\alpha, j_\beta) = p^\lambda(i_\alpha) \times p^\lambda(j_\beta). \tag{6.5}$$

(One might want to condition on the setting-events, M^α_i and M^β_j, but I take settings to be implied already by D^α_{i,i_α} and D^β_{j,j_β} — a detector cannot 'indicate' spin-up or spin-down in the d-direction unless the magnet was oriented in the d-direction. Moreover, I have qualms about including such events in a probability space, as I explain below.) Bell-factorizability is more or less the locality condition originally suggested by Bell, though adapted to a probabilistic setting.[1] I shall refer to it as 'factorizability' from now on. Any factorizable hidden-variables theory could be reasonably called 'local'. (When I use the term 'local' without qualification, I mean 'factorizable'.)

   In an attempt to understand factorizability, and to understand why and how quantum-mechanics violates it, some authors (notably Jarrett[2]) analyzed factorizability as the logical conjunction of two conditions. The statement of these two conditions requires a return to the description that I earlier rejected, and I shall discuss the significance of this fact below. The conditions are the following.[3]

Outcome independence: For any λ, i, j,

$$p^\lambda(D^\alpha_\pm, D^\beta_\pm | M^\alpha_i, M^\beta_j) = p^\lambda(D^\alpha_\pm | M^\alpha_i, M^\beta_j) \times p^\lambda(D^\beta_\pm | M^\alpha_i, M^\beta_j), \qquad (6.6)$$

where the ± 1 for $D^\alpha_\pm$ and the ± 1 for $D^\beta_\pm$ can be chosen independently of one another (i.e., (6.6) is really four conditions).

Parameter independence: For any λ, i, j,

$$\left.\begin{array}{l} p^\lambda(D^\alpha_\pm | M^\alpha_i, M^\beta_j) = p^\lambda(D^\alpha_\pm | M^\alpha_i), \\ p^\lambda(D^\beta_\pm | M^\alpha_i, M^\beta_j) = p^\lambda(D^\beta_\pm | M^\beta_j). \end{array}\right\} \qquad (6.7)$$

The logical conjunction of these conditions yields

Jarrett-factorizability: For any λ, i, j,

$$p^\lambda(D^\alpha_\pm, D^\beta_\pm | M^\alpha_i, M^\beta_j) = p^\lambda(D^\alpha_\pm | M^\alpha_i) \times p^\lambda(D^\beta_\pm | M^\beta_j). \qquad (6.8)$$

6.2.3 Understanding Jarrett-factorizability

Is Jarrett-factorizability, taken as the conjunction of outcome independence and parameter independence, physically interesting? To analyze factorizability as Jarrett does *requires* one to consider the 'outcomes' to be not the events $D^n_{d,\pm}$, but the events $D^n_\pm$. However, I argued above that the latter are experimentally empty. I do not mean, of course, that the bare notion of a detector's flashing is physically meaningless — it means just what it says, that the detector flashes. However, it is *experimentally* meaningless; the experiment is to measure the spin of a particle (or the polarization of a photon), and mere flashes of a detector without any extra information tell one nothing about the spin of particle (or the polarization of the photon). To put the point differently, we are interested in an analysis of the EPR–Bohm experiment, and it remains unclear what 'bare flashes' have to do with that experiment. Indeed, the usual consequences that are drawn from factorizability cannot be made in terms of bare flashes. For example, Bell's inequality (which I discuss in the next section) cannot be stated in terms of bare flashes, but requires reference to the settings of the detectors (or polarizers).

On the other hand, I am perhaps being too harsh on Jarrett-factorizability, because it does at least condition the probability of a 'bare flash' at a given detector on the orientation of that detector's corresponding magnet (or polarizer). Hence we might note that outcome independence and parameter independence are given in terms of these conditional probabilities, and claim that at least the conditional probabilities have some experimental significance.

I think that they do, but the question is whether the framework in which they are expressed is satisfactory. This framework countenances bare flashes as meaningful, informative events — that is, as outcomes of *measurements* — and for reasons already given, I doubt whether they are.

There are also reasons for worrying about a framework that takes the orientations of magnets or polarizers to be events in the probability space of a theory at all. (Note, however, that Jarrett himself does not fall foul of this worry. Rather than condition on orientations, he subscripts the probability measure with them.) The state of a magnet or polarizer is of course immensely complex, and the best we can do is to use the probabilities for various orientations given by a crude model. Perhaps *in principle* such probabilities exist in a complete theory (whether quantum mechanics or some other), but an analysis of locality that requires (even if only implicitly) that these probabilities can be specified cannot shake the nagging doubt that if the real story were known about how the orientations are chosen, the results of the analysis would not be different, or appear differently in the light of the details about how orientations are correlated with other events.

Hence I am skeptical of the significance of Jarrett's analysis because I am skeptical of the physical framework in which it is expressed — most especially, the analysis seems to me to rely on distinctions whose physical import is unclear, or anyhow, whose physical import can only rest on the details of a physical model of the experiment, the likes of which Jarrett's analysis (by virtue of its generality) cannot consider.

Finally, there is the question of what one is going to *do* with this analysis. Much has in fact been done with it, but all of this work — though valuable for what it teaches us — might in the end lead only to the conclusion that the distinction itself is not interesting, or useful (a conclusion which would itself, somewhat ironically, be quite interesting and useful). I will first consider and reject an argument of Maudlin's to this effect, then suggest that nonetheless, we may benefit by taking a point of view that ignores the distinction between outcome independence and parameter independence.[4]

Maudlin has pointed out that Jarrett-factorizability is equivalent not only to the logical conjunction of outcome independence and parameter independence, but also to the logical conjunction of the following conditions, stated here only in terms of α (but analogous conditions may, of course, be defined for β).[5]

Maudlin's outcome independence: For any λ, i, j,

$$p^{\lambda}(D_{\pm}^{\alpha}|M_i^{\alpha}, D_{\pm}^{\beta}) = p^{\lambda}(D_{\pm}^{\alpha}|M_i^{\alpha}). \tag{6.9}$$

Maudlin's parameter independence: For any λ, i, j

$$p^\lambda(D_\pm^\alpha | M_i^\alpha, M_j^\beta D_\pm^\beta) = p^\lambda(D_\pm^\alpha | M_i^\alpha, D_\pm^\beta). \tag{6.10}$$

It is clear that the mere existence of an alternative analysis does not render Jarrett's analysis useless. Jarrett's analysis might still be able to teach us something about the violation of locality. However, Maudlin suggests that indeed it cannot, for the following reason.

Standard quantum mechanics, says Maudlin, violates Maudlin's parameter independence, rather than Maudlin's outcome independence. On the other hand, standard quantum mechanics violates Jarrett's outcome independence, but not Jarrett's parameter independence. Therefore, we do not really learn that the non-locality of quantum mechanics is 'mediated' by outcomes rather than by parameters, as Jarrett and others suggest — *that* conclusion holds only in the context of Jarrett's analysis.

There is a minor flaw in this argument. Maudlin's claim that standard quantum mechanics violates his parameter independence condition but not his outcome independence condition holds only in the special case where the probability measure over possible measurements assigns equal weight to each direction of measurement. However, consider a case where, for example, the probability of M_j^β is much higher than the probability of $M_{j'}^\beta$, and suppose that $i = j$. Then Maudlin's outcome independence fails, because the probability that the detectors for α and β indicate opposite spins is very high — if spin is measured in the same direction on both sides then the detectors will indicate opposite spins.

However, the real problem with Maudlin's analysis, from the present point of view, seems rather to be that his conditions contain 'bare flashes'. If the usefulness of Jarrett's analysis is questionable because it is given (necessarily) in a *framework* that must admit bare flashes (and probability measures over apparatus-settings), Maudlin's is the more so because it actually countenances bare flashes — they appear in Maudlin's conditions unconditioned on apparatus-settings.

I do not mean to suggest, however, that I find Jarrett's analysis completely satisfactory, though it may be more satisfactory for some purposes than for others. There are two distinct problems to which Jarrett's analysis might be (and has been) applied. First, how are we to understand the metaphysics of non-locality? Are the non-local 'connections' in quantum mechanics best understood in terms of superluminal causation? In terms of the violation of some 'classical' metaphysical principle such as the spatio-temporal individuation of objects? Second, does the violation of non-locality, however it is understood, violate any part of the theory of relativity? In the end, Jarrett's analysis may be relevant to both questions, but it solves neither (though it has been taken by some to solve both).

Consider, for example, the question of whether there is superluminal causation between the two wings of the EPR–Bohm experiment. Some authors have suggested that the violation of outcome independence rather than parameter independence saves a theory from superluminal causation, and should be understood instead as a consequence of the fact that the particles are not ontologically distinct (despite their spatio-temporal separation). However, I contend that Jarrett's analysis is far too general to establish any such conclusion. Whether there is superluminal causation between the particles depends far less on the violation of outcome independence than on the details of the particular theory in question. Standard quantum mechanics might, for example, make plausible the idea that the particles are not distinct individuals, because it does not assign either of them its own statevector.[6] However, other equally satisfactory models of the EPR–Bohm experiment suggest otherwise. For example, the atomic version of the modal interpretation suggests that connection between the particles is due to a non-local dynamics, which is plausibly underwritten by superluminal causes.

Moreover, even if we allow that there is *some* superluminal causation between the wings, still the violation of outcome independence by itself should not be immediately taken to show that the causation is from outcome to outcome. First, it is not at all clear what 'outcome-to-outcome' causation would be, given that 'outcomes' (in Jarrett's sense) are not obviously physical events of any interest. Second, without the details of a model to hand, it is not clear that the causation might not be mediated in some other way. For example, there is a model of the experiment in which the violation of outcome independence *entails* the violation of parameter independence.[7] Moreover, in this model, the causes can be read — perhaps even plausibly so — as 'parameter-to-outcome' causes (ignoring, for the moment, the question of what these events mean physically).[8] Metaphysical questions about causation, individuation, and the like in the EPR–Bohm experiment simply cannot be plausibly decided at the very abstract level at which Jarrett's analysis occurs — one needs the details of a model to make plausible claims about the metaphysics underlying non-locality.

Similar statements hold about the relationship between non-locality and the theory of relativity and, in particular, Lorentz-invariance. Again, some authors have suggested that a theory can maintain Lorentz-invariance by upholding outcome independence and denying parameter independence. Again, whether this claim is true — and in fact I believe it to be in general *false* — must depend on the details of some model. In particular, the question of

Lorentz-invariance often turns on the dynamical details of a model, a fact to which I will return in the next two chapters.

Even questions about signalling — which were largely Jarrett's original concern — are not finally answered by Jarrett's analysis. It is sometimes supposed that parameter-independence was *proven* by Jarrett to be equivalent to the possibility of superluminal signalling. However, this view of Jarrett's argument is quite misleading. As Jarrett acknowledges, a failure of parameter-independence permits signalling *only* if the 'hidden' variables, λ, can be controlled by the signaller. If the λ cannot be controlled, then the best one can say is that signalling is *possible*. Is the mere physical possibility (but not the actuality) of signalling of any interest? In particular, does it conflict at all with the theory of relativity? Later I will suggest that it might not. Moreover, it might happen — and indeed, we have already seen that it *does* happen in Bohm's theory — that the λ cannot, even in principle, be controlled. It is *physically impossible* in Bohm's theory to control the λ; but then in what sense is signaling at all possible in Bohm's theory?

Hence I am skeptical that Jarrett's analysis can, by itself, teach us much about non-locality. I do not mean to say that Jarrett's analysis should be dropped — though in fact I am going to drop it here in favor of considering just Bell's locality condition — but rather that it should always be applied in the context of a given (detailed) model of the EPR–Bohm experiment. Even then, its application may point to conclusions whose physical meaning is unclear, due to its reliance on the events $D^n_\pm$ and M^n_d, whose physical import is unclear.

6.3 Bell's theorem

Bell used his locality condition to prove the most significant theorem to date on non-locality in quantum mechanics.[9] Bell considered hidden-variable theories of the sort mentioned here. He pointed out that to recover the predictions of quantum mechanics, the λ must be distributed according to a function, $\rho(\lambda)$, such that the following condition holds:

Empirical adequacy for hidden-variables theories: For any λ, $d = i, j$, $n = \alpha, \beta$, and $d_n = \pm 1$,

$$\int_\Lambda p^\lambda(d_n)\, \rho(\lambda)\, d\lambda = p^\psi(d_n), \tag{6.11}$$

where $|\psi\rangle$ is the quantum-mechanical state of the pair of particles at the time of measurement, and Λ is the set of all λ.

In other words, when we average over our ignorance about the actual complete state, λ, of a system, we should recover the quantum-mechanical predictions. A strictly analogous condition should hold for the joint probabilities when quantum mechanics delivers joint probabilities.

Bell was originally concerned with only theories in which λ fixes the outcome of the measurement. As Bell did, I take the λ to be the complete state 'at some suitable instant' prior to measurement — perhaps when the particles are created, or perhaps just before measurement. In either case, the kind of 'determinism' involved is not fully dynamical yet, but is instead concerned with only two times, the 'suitable instant' described by λ and the time of measurement. Hence I use the name 'two-time determinism'.

Two-time determinism: For any λ, d, n, and d_n

$$p^\lambda(d_n) = 0 \text{ or } 1. \tag{6.12}$$

However, later Bell considered two-time stochastic theories as well (i.e., ones for which two-time determinism fails).

In both the deterministic and the stochastic case, Bell required

λ**-independence**: The distribution $\rho(\lambda)$ is independent of the setting of the apparatus.

Bell then proved:

Bell's theorem: Any factorizable, λ-independent, empirically adequate model of the EPR–Bohm experiment is committed to Bell's inequality.

Bell's inequality is violated by quantum mechanics. Hence no factorizable, λ-independent, empirically adequate model of quantum mechanics exists.[10]

6.4 Determinism and factorizability

6.4.1 Two-time determinism and factorizability

It was shown by Suppes and Zanotti that a cousin of two-time determinism follows from outcome independence. However, as one might expect, the events used to formulate this cousin of two-time determinism are of the sort that I rejected earlier — the same sort that are used to formulate outcome independence itself. Here I shall show how the derivation of Suppes and Zanotti goes through for Bell-factorizability rather than outcome independence.[11]

The proof uses the strict correlations of quantum mechanics. As we have seen, whenever i and j are parallel or anti-parallel — i.e., whenever $\theta_{ij} = k\pi$ for any integer, k — quantum mechanics yields strict correlations such as the following:

Strict correlations: For the singlet state, $|\psi\rangle$ (see (6.1)), and any $i = j$,

$$p^{\psi}(i_{\alpha} = j_{\beta}) = 0. \tag{6.13}$$

Empirical adequacy for the joint probabilities therefore requires that for 'measure-1' (where ρ is the measure — this measure arises from the application of empirical adequacy to recover the quantum-mechanical strict correlations[12]) of the λ, and for any $i = j$:

$$p^{\lambda}(i_{\alpha} = j_{\beta}) = 0. \tag{6.14}$$

By factorizability, it follows that for any $r = \pm 1$

$$p^{\lambda}(i_{\alpha} = r) = 0 \quad \text{or} \quad p^{\lambda}(j_{\beta} = r) = 0. \tag{6.15}$$

Now, if we assume that for any n and any d

$$p^{\lambda}(d_{n} = r) = 1 - p^{\lambda}(d_{n} = -r) \tag{6.16}$$

(which is just the usual condition of normalization on probabilities), then it follows from (6.15) that for any n, d, and r:

$$p^{\lambda}(d_{n} = r) = 0 \text{ or } 1, \tag{6.17}$$

which is just two-time determinism. Therefore, the strict correlations of quantum mechanics plus factorizability (plus the condition of empirical adequacy and the sum-to-one condition) together entail that two-time determinism holds for at least measure-1 of the λ.

The intuitive reason is that when the magnetic fields are perfectly aligned (parallel or anti-parallel), the quantum-mechanical probabilities are 0 or 1. In general, one could allow that when the magnetic fields are not perfectly aligned (and therefore the quantum-mechanical probabilities are not 0 or 1), the probabilities of the hidden-variable theory are also not 0 or 1, but factorizability entails that the probabilities of the hidden-variable theory for the result at one region are insensitive to whether the magnetic field in that region is aligned with the magnetic field of another region. Therefore, in order to be adequate for the cases when the fields *are* aligned, the probabilities for the result at one region must *always* be 0 or 1.

One might suspect that this connection between factorizability and two-time determinism is an accident of the two-particle case, but in fact it is not. The connection is a conceptual one — i.e., it does not depend on the two-particle case, but only on the existence of strict correlations. Indeed, the intuitive argument for two-time determinism in the case of N particles goes just like the argument of the previous paragraph for the case of two particles.[13] Of course, it must be admitted that we leave experiment behind as soon as we move away from the two-particle case — EPR–Bohm type experiments have not been performed on systems of more than two particles — but the point is that there is a conceptual connection between factorizability and two-time determinism in hidden-variables theories.

So factorizability entails two-time determinism. The converse is also true: two-time determinism entails factorizability. The reason is simple: if an event has probability 0 or 1, then conditionalization on any other event does not change its probability. (This fact makes intuitive sense, and follows immediately from the definition of conditional probability.) Hence, for example, if $p^\lambda(i_\alpha = +1) = 0$ then

$$p^\lambda(i_\alpha = +1 | j_\beta = +1) = 0 = p^\lambda(i_\alpha = +1) \qquad (6.18)$$

and it is a short step from equalities of this sort to factorizability. Therefore, given the strict correlations of quantum mechanics, factorizability and two-time determinism are equivalent.

6.4.2 Model determinism and factorizability

The strict correlations of quantum mechanics clearly played an important role in the derivation of two-time determinism from factorizability. One might wonder how far we can go towards establishing a link between determinism and factorizability *without* using the strict correlations.

The answer is: remarkably far. Fine showed how.[14] He begins by countenancing two types of model of the EPR–Bohm experiment: deterministic factorizable models and stochastic factorizable models. The former obey the condition of two-time determinism and the latter do not. Both are presumed empirically adequate. Now, we know from the previous subsection that the latter sort of model is, in fact, impossible — there are no two-time stochastic, factorizable, empirically adequate models of the EPR–Bohm experiment. However, to make this objection is to miss the point of this exercise, which is to see how far we can go towards deriving determinism *without* making use of the strict correlations of quantum mechanics.

For my purposes, Fine's result[15] is best stated in terms of the following forms of determinism.

Model determinism: A model of the EPR–Bohm experiment is 'model deterministic' if it admits an empirically equivalent model that is two-time deterministic.

Factorizable model determinism: A model of the EPR–Bohm experiment is 'factorizably model deterministic' if it admits an empirically equivalent factorizable model that is two-time deterministic.

In the present context, 'empirical equivalence' means 'agreeing on the probabilities of the empirically relevant events' (i.e., the events expressible in the language of quantum mechanics — in our case, the $D_{d,\pm}^n$). Fine proved that every stochastic factorizable model for the EPR–Bohm experiment meets the condition of factorizable model determinism.

Be careful, however, not to draw the following fallacious inference: Fine showed that there exists a stochastic factorizable model if and only if there exists a deterministic one; but by the results of the previous subsection (due to Suppes and Zanotti) there does not exist a stochastic factorizable model, and therefore, there does not exist a deterministic one. The fallacy is that what Fine showed is that there exists *some* stochastic factorizable model if and only if there exists a deterministic one, but what the result of the previous subsection shows not to exist is a *genuinely* stochastic factorizable model. There is still room for a 'trivially' stochastic factorizable model, i.e., one that is also deterministic. Bell's theorem is needed to rule these out.

Fine's proof makes no particular assumptions about the quantum-mechanical state of the pair of particles, nor about whether strict correlations hold. Hence, as advertised, Fine's result establishes a generic link between factorizability and determinism — any factorizable model *must* obey factorizable model determinism. From this (and related) results, Fine concluded that

despite appearances, no significant generality is achieved in moving from deterministic HV-[hidden-variable]models to stochastic ones, if factorizability is required of the latter.[16]

Model determinism is evidently weaker than two-time determinism, and some authors have used this apparent weakness to question Fine's claim that no generality is had by considering stochastic models. Their claim is based on a distinction between the 'physically real' probabilities, and the 'mathematically possible' probabilities.[17]

For example, Butterfield acknowledges the vagueness of the term 'physically real' as applied to probabilities, but makes the following minimal requirements:[18]

(i) Physical reality requires something more than just successfully modelling the given statistics.
(ii) There is at most one physically real probability distribution on the quantities.

It follows immediately that some genuinely stochastic theory could give the 'physically real' probabilities, while the factorizable deterministic model (guaranteed to exist by Fine's result) is 'merely mathematical'. Butterfield's main contention, then, is that while perhaps no *mathematical* generality is achieved by the move from deterministic to stochastic models, *physical* generality might be achieved.

To assess this reaction to Fine, first distinguish two interpretations of Fine's claim that no generality is achieved by moving to stochastic models. On the one hand, his claim might be the weak one that stochastic models are committed to Bell's inequality if and only if deterministic models are. Recall that Bell's original derivation of the inequality was restricted to deterministic models. Fine's weak claim is that this original derivation already implicity covered the case of stochastic models. On the other hand, Fine's claim might be interpreted as the strong one that every purportedly stochastic factorizable model is, in fact, deterministic.

The weak claim is true, and is so by virtue of Fine's result. The existence of a stochastic factorizable model entails the existence of a deterministic factorizable model, and the existence of a deterministic factorizable model entails Bell's inequality (assuming λ-independence, as noted earlier — but, of course, λ-independence is needed to get Bell's inequality for a stochastic model as well). Note that whether one considers the latter model to be 'physically real' is irrelevant — Bell's theorem says not that Bell's inequality follows from the existence of a physically real deterministic factorizable model, but that it follows from the existence of *any* deterministic factorizable model.

The weak claim is also not very weak. One might suspect that it is, because it is always possible to construct a deterministic model from a stochastic one, for the sample space of the stochastic theory may always be considered a subspace of a larger sample space, each of whose elements assigns probability 0 or 1 to the empirical events modelled by the original stochastic model. However, features of the original stochastic model will not necessarily carry over to the more fine-grained deterministic model. In particular, the factorizability of the stochastic model will not necessarily translate into factorizability in the more fine-grained deterministic model. One achievement of Fine's was to show that in the EPR–Bohm experiment, one can always use a factorizable stochastic model to construct a factorizable deterministic model.

The strong claim, if true, would nullify the criticism based on physical reality, for it would show that the scenario envisaged in this criticism — namely, one in which a genuinely stochastic and 'physically real' factorizable model exists — is impossible. However, it does not appear that the strong claim follows immediately from Fine's results, which, as I noted, show only how to construct a deterministic factorizable model from a stochastic one.

However, the strong claim *is* true. As Butterfield himself acknowledges,[19] it follows from the result of the previous subsection. In other words, despite appearances, there is, in fact, *no such thing* as a genuinely stochastic factorizable hidden-variable model of the EPR–Bohm experiment. However, this claim requires us again to use the strict correlations, and therefore it is worth asking what plausibility there is in denying their use.

The answer is: not much. There is the obvious fact that quantum mechanics predicts them; therefore any theory that relies on the *absence* of strict correlations will not be an interpretation of quantum mechanics, but instead a new theory (and one very likely to produce significant (i.e., detectable) empirical disagreement with quantum mechanics). The more serious problem, however, is that these correlations are no accident of the formalism of quantum mechanics. They are direct manifestations of the conservation of spin, the denial of which is more serious than the supposition that quantum mechanics is not quite right.

Fine's result strengthens these points, for it shows that if one wants to argue that the 'true theory' of the world is factorizable and stochastic, then not only must one deny the strict correlations, but also one must be prepared to argue that the factorizable deterministic model that Fine's result guarantees to exist is not 'physically real'. Probably any such argument will not be overwhelmingly convincing.

6.5 Can there be a local model?

We may conclude that any factorizable theory must be two-time deterministic, and that any two-time deterministic theory must be factorizable. However, factorizability leads to Bell's inequality. Must we therefore conclude that two-time deterministic theories are unacceptable? For example, given the appropriate identification of the λ (as I will discuss in chapter 9), Bohm's theory is two-time deterministic, and some authors have suggested that Bell's theorem rules out Bohm's theory.[20] Are they right?

The answer is: no. Recall that λ-independence is also required for the proof of Bell's theorem, and Bohm's theory explicitly denies λ-independence. I will discuss this point in detail in chapter 9.

Hence there appears to be room for a factorizable theory that does *not* violate the Bell's inequality. Moreover, if there is room for such a theory, then *only* a two-time deterministic theory can take advantage of it. On the other hand, stochastic theories *must* violate factorizability, because there are no factorizable two-time stochastic theories. If we hope for a local theory, then we are forced to consider deterministic theories.

7

Non-locality II: Dynamical models of the EPR–Bohm experiment

7.1 Dynamical determinism

7.1.1 Dynamical models of the EPR–Bohm experiment

In the previous chapter, I considered two-time models of the EPR–Bohm experiment, which are simplifications of more realistic models, namely, those providing a complete dynamics. In this chapter, I consider determinism and locality in such models. Again I restrict attention to the EPR–Bohm experiment, though it should be remembered that non-locality is a generic feature of quantum mechanics, not restricted to a few special experiments.

In a fully dynamical model of the EPR–Bohm experiment, we may associate with the emission of the pair of particles an initial time, t_0. (In this chapter I shall consider only the two-particle case.) With the outcome of a measurement on particle n, we may associate a later time, $t_n > t_0$. The outcomes may be written:

$D^n_{d,\pm}(t^n) \equiv$ the event: particle n's detector indicates spin-$\pm\hbar/2$ in the d-direction at time t^n.

The time, t^n, is not meant to suggest that the projections representing these events are time-dependent (though we could make them so by working in the Heisenberg picture), but only that the event occurred at time t^n. I shall not require that t^α and t^β be equal, though they may be. (In a relativistic setting, discussed later, this requirement is in any case meaningless.) As in chapter 6, where possible I use the simpler notation, i^t_α, where $i^t_\alpha = \pm 1$ if and only if $D^\alpha_{i,\pm}(t^\alpha)$ occurs, and similarly for j^t_β.

For simplicity (and without loss of generality), we may assume that the evolution operator taking the quantum system from the initial state at t_0 to the state just prior to measurement is the identity, so that the initial wave function, the singlet state, may be used to calculate the quantum-mechanical

147

probabilities for the possible results. Hence the quantum-mechanical probabilities are written as before: $p^\psi(\cdot)$, where $|\psi\rangle$ is the initial quantum-mechanical state (and therefore also the final state) of the pair of particles. For us, $|\psi\rangle$ is usually the singlet state (6.1).

The notation above avoids mention of the times at which the magnets are oriented, and does so by design. As I explained in chapter 6, I prefer an analysis of the experiment that does not make mention of such events. (The relevant information that we would get from such events is already included in the events $D^n_{d,\pm}(t^n)$, through the value of d.) Indeed, the move to a dynamical picture changes nothing in my argument that the outcomes of the experiment are best represented as $D^n_{d,\pm}$, and I shall therefore take as given that in the dynamical case the outcomes are best represented as $D^n_{d,\pm}(t^n)$.

Because we have introduced time into the description of the experiment, we need also to introduce it into our notion of a hidden-variables model. To do so, represent the complete state of the pair of particles as a stochastic process, $L(t)$. (A stochastic process is just a time-indexed family of random variables.[1] There may or may not be non-trivial correlations between the values of the random variables at different times.) This process takes a value, $\lambda \in \Lambda$, at each time, t. I assume that any adequate dynamical model would provide transition probabilities for $L(t)$. The probability that the complete state at t' is λ', given that it is λ at t ($t < t'$) is such a transition probability, and may be denoted

$$p\Big(L(t') = \lambda'|L(t) = \lambda\Big) =_{\mathrm{df}} p(\lambda'|\lambda). \tag{7.1}$$

I will use the right-hand side of (7.1) whenever the meaning is clear. Further, I assume that $p\Big(L(t') = \lambda'|L(t) = \lambda\Big)$ is defined for all $t' \geq t$, λ, and λ'.

We may then extend the notation used in chapter 6 for single-time probabilities to give probabilities for various outcomes of the experiment, given the initial complete state of the particles. As before, I take these probabilities not to be probabilities for various d^t_n conditional on various initial states, λ_0. I suppose merely that initial states generate probability measures over results as follows:

$$p^{L(t_0)=\lambda_0}(d^t_n) =_{\mathrm{df}} \int_\Lambda p^{\lambda_{t^n}}(d^t_n) \times \rho(\lambda_{t^n}|\lambda_0)\, d\lambda_{t^n}, \tag{7.2}$$

where λ_{t^n} here is the complete state at time t^n and $\rho(\lambda_{t^n}|\lambda_0)$ is the transition probability density obtained from $p(\lambda_{t^n}|\lambda_0)$ in the usual way.[2] When no ambiguity results, I use the following notation:

$$p^{L(t_0)=\lambda_0}(d^t_n) =_{\mathrm{df}} p^{\lambda_0}(d^t_n).$$

(The danger in this notation is that it might lead one to think that $p^{\lambda_0}(d_n^t)$ is a single-time probability. It is not. The clue is that λ_0 is the state at time t_0 whereas $D_{d,d_n^t}^n(t^n)$ occurs at time t^n.)

In an adequate dynamical model, the transition probabilities obey a condition of empirical adequacy:

Empirical adequacy for transition probabilities: For any λ, d, n, and d_n^t

$$\int_\Lambda \int_\Lambda p^{\lambda_t}(d_n^t) \times \rho(\lambda_n|\lambda_0) \times \rho(\lambda_0)\, d\lambda_n\, d\lambda_0 = p^\psi(d_n^t), \tag{7.3}$$

where $|\psi\rangle$ is the quantum-mechanical state of the pair of particles at t_0 (and therefore, by assumption, at t^n as well). Roughly, (7.3) takes each possible initial state, λ_0, calculates the probability of the event $D_{d,d_n^t}^n(t^n)$ given λ_0, multiplies the result with the probability of beginning in λ_0 in the first place, and sums (integrates) over all such weighted results (i.e., one for each value of λ_0).

7.1.2 Two kinds of dynamical determinism

In what ways could a dynamical model of the EPR–Bohm experiment be deterministic? Two are immediately evident.

The first is that the transition probabilities could be deterministic:

Deterministic transitions: For any $t < t'$, any λ, and any λ',

$$p(\lambda'|\lambda) = 0 \text{ or } 1. \tag{7.4}$$

In this case, given any initial state $L(t_0)$, $L(t)$ is just a (deterministic) function of time.

The second is that the initial state of the particles could determine the results of the experiment:

Deterministic results: For any $t_0 < t^n$, n, d, d_n^t, and λ_0,

$$p^{\lambda_0}(d_n^t) = 0 \text{ or } 1. \tag{7.5}$$

This second form of determinism requires two comments.

First, if it holds, it holds for *every* $t^n > t_0$, where t^n is considered a time at which the outcome of the experiment occurs. However, it is possible that, given the complete initial state λ_0, the time of the occurrence of the outcome is fixed. (Exactly this situation occurs in Bohm's theory under a suitable definition of λ_0, as I shall discuss in chapter 9.) In such cases, one might think it best to consider $p^{\lambda_0}(d_n^t)$ to be undefined for all values of t^n except the time at which the outcome will occur. The condition of deterministic results could be modified to take this case into account, but it need not be, for it is equally reasonable to set $p^{\lambda_0}(d_n^t)$ equal to zero when t^n is not the time at which the outcome will occur.[3]

Second, the condition does not cover the case where $t_0 = t^n$. The reason is that the condition of deterministic results is meant to capture a form of *dynamical* determinism, whereas the case $t_0 = t^n$ is not dynamical. The condition one gets in the case $t_0 = t^n$ is really a condition on the nature of the λ, saying that each λ is associated with one and only one outcome of the experiment:

λ-**determination**: Given that the time of the occurrence of the outcome for particle n is t^n, the complete state at t^n, λ_{t^n}, completely determines the outcome.

This condition might appear strong at first — for example, it might appear to be an assumption of determinism — but, in fact, it will turn out to be trivial on my definition of the λ. (They will turn out to be complete states of regions of space-time, including the region occupied by the detector.) Indeed, any account that allows λ-determination to fail would seem to be quite inadequate. How can we call the λ 'complete' states when in fact they do not distinguish states that are manifestly distinct, i.e., states of different spin? (This point is especially vivid if we allow the λ to be the complete states of the entire system, composed of the particles and the apparatuses.) Therefore, I assume that λ-determination holds regardless of whether deterministic results does.

The conditions of deterministic transitions and deterministic results are not entirely independent. The obvious connection is that the former condition entails the latter. The intuitive reason is that, given any λ_0, the condition of deterministic transitions fixes the state, λ_n, at t^n — call this state 'λ_n^*' — and the assumption of λ-determination translates this inevitable λ_n^* into an inevitable result for the measurement. The mathematical reason is evident in (7.2), where $\rho(\lambda_n|\lambda_0)$ becomes a delta function, $\delta(\lambda_n^* - \lambda_n)$, taking the integral to $p^{\lambda_n^*}(d_n^t)$ which, by λ-determination, is 0 or 1.

However, it is more useful to see why the reverse implication does not hold, i.e., why the condition of deterministic results (plus λ-determination) does *not* entail deterministic transitions. To do so, assume deterministic results. For a given $t_0 < t^n$, n, d, d_n^t, and λ_0, consider the following two cases (the only ones possible, by deterministic results). First, $p^{\lambda_0}(d_n^t = r_n) = 0$. Looking back to (7.2) we see that it is sufficient for this case to hold that $\rho(\lambda_n|\lambda_0) = 0$ whenever $p^{\lambda_n}(d_n^t) \neq 0$ and vice versa. Second, $p^{\lambda_0}(d_n^t) = 1$. In this case, it could be that $p^{\lambda_n}(d_n^t) = 1$ for more than one value of λ_n, but that all transitions from λ_0 are to some such λ_n.

The possibility that the conditions of deterministic results and λ-determination hold, while the condition of deterministic transitions fails, illustrates an important point. One might well wonder whether the question 'Is *the world* deterministic?' has any hope of being answered, and whether it is useful to ask in the first place. Even if, somehow, we could establish that some model of the EPR–Bohm experiment is deterministic in the sense of exhibiting deterministic results and λ-determination, and that it is 'physically true' (whatever that means), the possibility of an underlying indeterminism (expressed by the denial of deterministic transitions) remains. If we can get hold of the λ empirically to test deterministic transitions, who can say that there are not more fine-grained states that evolve indeterministically, while preserving the observed determinism at the level of the λ? Of course, the argument goes the other way as well: claims on behalf of quantum mechanics notwithstanding, the probabilities of any indeterministic theory can, of course, be merely epistemic.

However, the possibility of underlying determinism or indeterminism does not put an end to all discussion or investigation. In particular, we may build the possibility of an underlying indeterminism into the definition of deterministic transitions, by coarse-graining the complete states. Let us suppose that in some model, we can find equivalence classes, Λ_k^t, at each time t such that, *given* the equivalence class containing the complete state at the initial time, we can predict with certainty (for each later time) which equivalence class will contain the complete state. Such a theory obeys the following condition:

Weak deterministic transitions: For any $t_0 < t \leq t^n$, any equivalence class Λ_k^t at time t, and any initial state λ_0,

$$p\left(L(t) \in \Lambda_k^t | L(t_0) = \lambda_0\right) = 0 \text{ or } 1. \tag{7.6}$$

The possibility of an indeterministic theory underlying a deterministic one may now be expressed as the possibility that weak deterministic transitions holds, while deterministic transitions fails. (Note also that the latter entails the former, because we can take the equivalence classes at times other than measurement to be just the individual λ themselves.)

Of course, every dynamical model is weakly deterministic in this sense, where we take $\Lambda_k^t = \Lambda$ for all t. However, we are interested in non-trivial examples of weakly deterministic theories, for which Λ_k^t is a proper subset of Λ for at least some t. Moreover, we are interested in weakly deterministic theories whose equivalence classes at the time of measurement, t^n, are exactly the equivalence classes one would get by using λ-determination to partition the set of all complete states into one equivalence class for each possible result of the experiment. We could write

$$p^{\lambda_n}(d_n) = \begin{cases} 1 & \text{if } \lambda_n \in \Lambda_{d_n} \\ 0 & \text{if } \lambda_n \notin \Lambda_{d_n}. \end{cases} \tag{7.7}$$

Models that are weakly deterministic in *this* way I shall call 'empirically deterministic'.

Weak determinism is meant to be completely compatible with contextualism. In particular, it requires only that *for a given d and t_n*, the model satisfy weak determinism for the equivalence classes Λ_{d_n} together with non-trivial equivalence classes for earlier times. It does *not* require that at time t_n classes Λ_{d_n} are simultaneously definable for all d.

The condition of *d*deterministic results, plus λ-determination, entails empirical determinism. Again, the reason is intuitively evident. The assumption of λ-determination guarantees the existence of equivalence classes at the time of measurement, while the condition of deterministic results guarantees that at every possible time of measurement, there is some equivalence class in which $L(t)$ is guaranteed to be. (To be clear on what is meant here, contrast this result with the obvious fact that $L(t)$ is guaranteed to be in some equivalence class at each time, because it is single-valued.) To see the reason mathematically, use conservation of probability along with deterministic results to find for any t^n the value of d_n^t such that

$$p^{\lambda_0}(d_n^t) = 1. \tag{7.8}$$

Then, by definition

$$\int_\Lambda p^{\lambda_n}(d_n^t) \times p(\lambda_n|\lambda_0)\, d\lambda_n = 1. \tag{7.9}$$

By λ-determination, $p^{\lambda_n}(d_n^t) = 0$ or 1, and may therefore be written as a characteristic function [4] of the equivalence class Λ_{d_n}, which reduces the integral in (7.9) to

$$\int_{\Lambda_{d_n}} p(\lambda_n|\lambda_0)\, d\lambda_n = 1 \tag{7.10}$$

and therefore, because $p(\lambda_n|\lambda_0)$ is normalized (its integral over all λ_n is 1), weak deterministic transitions must hold, to within a set of λ_n of measure zero, the measure being given by $p(\lambda_n|\lambda_0)$. That is, for each t^n, there is some equivalence class Λ_{d_n} such that the probability that $L(t)$ does not evolve from its initial state, λ_0, to some $\lambda \in \Lambda_{d_n}$ is zero.

There is one subtlety in this result: namely, the assumption that the condition of deterministic results, plus λ-determination, gives an equivalence class at each possible time of measurement. As I mentioned before, there may be only one such time, given the initial state, λ_0. However, I shall not discuss this case further — if the *time* of measurement is fixed by the initial state (as it is in Bohm's theory, for example) then very likely we are dealing with a fully deterministic theory (as is Bohm's theory), in which case weak deterministic transitions holds because deterministic transitions holds.

7.2 Dynamical locality

7.2.1 Dynamical factorizability?

Now that we have at least a preliminary understanding of determinism in dynamical models, how shall we formulate a locality condition that is appropriate to dynamical models? Perhaps the obvious place to begin is with a slightly more general version of factorizability as given in chapter 6:

Factorizability with dynamics: For any λ_0, t_0, t^α, t^β ($t_0 < t^\alpha, t^\beta$), i, and j:

$$p^{\lambda_0}(i_\alpha^t, j_\beta^t) = p^{\lambda_0}(i_\alpha^t) \times p^{\lambda_0}(j_\beta^t). \tag{7.11}$$

However, this condition is clearly not an acceptable locality condition. For example, the fact that $D_{i,+}^\alpha(t^\alpha)$ occurs could provide information about how $L(t)$ evolved, thus leading to information about the likelihood of the occurrence of $D_{i,+}^\alpha(t^\alpha)$. Intuitively, there is no non-locality in such a scenario. We are led to suspect that factorizability as given in chapter 6 is not a suitable locality condition for a fully dynamical model of the EPR–Bohm experiment.[5]

This suspicion is deepened by the following example. Suppose that the process $L(t)$ is a 'slippery slope' process, in which the state at the time of measurement is effectively decided soon after the emission of the particles, while they are still *not* space-like separated (though at the time of emission, the states of the particles could evolve in several ways). In that case, the occurrence of $D_{i,+}^\alpha(t^\alpha)$ at one wing of the experiment provides information about the early history (just after émission) of the evolution of $L(t)$, and therefore about the likelihood of various outcomes for β.

However, if the particles leave the source at the speed of light, in opposite directions, then surely they *do* evolve independently, in a local theory, and in the experiment with photons at least, they do travel at the speed of light. Perhaps factorizability with dynamics is a reasonable locality condition in such cases. However, in this case the condition that it really expresses is apparently the independence of the evolution of the particles. It is better to formulate such a condition explicitly. Hence, rather than seeking an adequate formulation of factorizability, I shall concentrate on formulating a condition of independence of the evolution of the particles.

7.2.2 *Digression: On the separability of physical objects*

To formulate a condition of independence of the evolution of the particles, note first that I have been implicitly assuming that it is possible in the first place to speak of the evolution of a single particle, without mentioning the other particle. Some would call this assumption already a 'locality condition', but the term is, at least in the present discussion, a misnomer.

On the other hand, this assumption is important enough to warrant a clarification in the form of a short historical digression. In a letter to Max Born, Einstein wrote:

It is ... characteristic of ... physical objects that they are thought of as arranged in a space-time continuum. An essential aspect of this arrangement of things in physics is that they lay claim, at a certain time, to an existence independent of one another, provided these objects "are situated in different parts of space". Unless one makes this kind of assumption about the independence of the existence (the "being-thus") of objects which are far apart from one another in space — which stems in the first place from everyday thinking — physical thinking in the familiar sense would not be possible. It is also hard to see any way of formulating and testing the laws of physics unless one makes a clear distinction of this kind. This principle has been carried to extremes in the field theory by localizing the elementary objects on which it is based and which exist independently of each other, as well as the elementary laws which have been postulated for it, in the infinitely small (four-dimensional) elements of space.

The following idea characterizes the relative independence of objects far apart in space (A and B): external influence on A has no direct influence on B; this is known as the "principle of contiguity," which is used consistently in the field theory. If this axiom were to be completely abolished, the idea of the existence of (quasi-)enclosed systems, and thereby the postulation of laws which can be checked empirically in the accepted sense, would become impossible.[6]

Einstein's last statement — that an external influence on A has 'no direct influence' on B — might lead one to believe that he is espousing a locality principle of the type often found in discussions of the EPR–Bohm experiment, and discussed in chapter 6. That is, he might be saying that events at A and B are probabilistically independent, in the sense of each being screened off from the other by its own backwards light cone (in the case of the EPR–Bohm experiment, by the initial state of the pair of particles). However, in light of the preceding paragraph of his letter, I prefer *not* to interpret Einstein thus.

Instead, I think Einstein is making a much more fundamental claim: that the possibility of science requires the possibility of describing a system as being located in a continuously connected spatio-temporal region. In this case, the phrase 'no direct influence' may mean *not* that there are no non-local forces, but that an influence in one region of space-time cannot be considered a 'direct' influence on a space-like separated region — 'direct' here meaning 'direct as a consequence of the fact that one and the same system (directly influenced as a whole) occupies both regions'. It may be only a denial of holism, and not of non-local action. Of course, Einstein was *also* not fond of non-local action, but I think that his primary worry in this passage is holism, not action at a distance. (On the other hand, one might argue that Einstein simply did not make a distinction between 'holism' and 'non-local action'. In that case, I am here emphasizing one aspect of what was for Einstein a more general concept of 'locality', comprising and not distinguishing the denial of holism *and* the denial of non-local action.[7])

If I am right about Einstein (though the historical question is not primary here), then he was making a powerful and, to me, persuasive claim. The claim is that the meaningfulness of experimental procedures depends on the meaningfulness of descriptions of physical systems as occupying continuously connected regions of space-time. Put differently, the claim is that the *state* of every physical object refers to, or is the state of, a single continuously connected region of space-time, the region that the object may be said to 'occupy'.

One might urge against this claim that two systems can be 'approximately isolated', even if, in reality, they are 'parts' of a single system occupying a region of space-time that is not continuously connected. However, this objection misses the point, which is not about approximations. Newtonian gravity provides an example. We may say, with some degree of rigor, that classical gravitational effects on experimental apparatuses are (usually) 'negligible', but saying so already presupposes that there are apparatuses to speak about in the first place, i.e., that there are localized objects upon which gravity may act. There is no sense — at least no established sense — in which we may say with rigor that an apparatus is 'mainly' located 'here'.

There is a weaker version of Einstein's claim: namely, that the state of a single particle can be completely characterized without reference to any other. This version of the claim is even more compelling. If the apparatus, i.e., the thing to which we think we refer in our everyday use of the word 'apparatus', is not an *object*, if it is essentially and irremovably (even in thought) a 'part of'[8] something else, then how may we be said to do experiments with *it*, the 'apparatus' (which is not really a thing by itself at all)? Again, any talk of approximations will not seem to help. To be able to say that 'this thing here approximates the whole' we need already the notion of 'this thing here'; but it is exactly this notion that holism threatens.

Some authors[9] have suggested that holism of some type might explain the apparent non-locality of quantum mechanics. Partly for reasons already given, I cannot see how holism is even a tenable scientific doctrine, much less an explanatory one. The basic idea of the proposal is clear: the 'two' particles are really one object, and therefore no matter how far 'apart' 'they' are, an influence on 'one' of them is an influence on 'both', and need not be thought to travel superluminally 'from one to the other'.[10] However, apart from the fact that it is not clear how such a scenario is any better off with respect to the theory of special relativity, it is not clear how one is to keep the disease from spreading. Why are our apparatuses not also 'parts of' holistic objects? Indeed, it seems very likely that they would be. But then our troubles are not merely semantic — it is not merely that we must invent a language for speaking about holistic objects, but that the objects that we would end up speaking about are *not* the ones with which we thought we were doing experiments. Indeed, *those* things (the things with which we thought we were doing experiments) are not objects at all — or at any rate, they are not in the condition that we thought they were in, namely, 'located here'. But then in what sense and with what objects *have* we done our experiment? And how is the experiment that we really did related to the one we thought we did? Holism raises these and still other difficult questions, and at what gain? It seems that, whether we call the non-locality 'action between objects at a distance' or 'correlations between two "parts" of the same object', we have the same trouble, namely, correlations of space-like separated events. The last-ditch maneuver by holists would have to be to claim that 'two' events are not involved here (and hence 'they' are not space-like separated. However, that claim would require radical revision in the theory of relativity, which is founded on the notion of a manifold of distinct events, and thus far there has been little indication how such a revision would go.

In any case, I shall adopt Einstein's claim, as I interpret it. Following Howard, I call it 'separability'. However, I do not follow Howard in supposing that separability entails anything about locality.[11] Separability as I conceive it is compatible with non-local interactions of all sorts. The only condition is that it be possible to speak of these interactions as non-local interactions between or among distinct systems.

Separability can be made precise in terms of the stochastic process, $L(t)$:

Separability: The complete state of the pair of particles in the EPR–Bohm experiment is given by $L(t)$, and

$$L(t) = \Big(L_\alpha(t), L_\beta(t) \Big) \tag{7.12}$$

is a random vector for each t. $L_n(t)$ is the complete state of particle n.

Calling $L(t)$ a random vector is a matter of convenience here. Technically, the $L_n(t)$ need not be assumed to be scalar-valued objects, but could be vector-valued, matrix-valued, or whatever. (See note 1.) The point is simply that, whatever the mathematical nature of $L(t)$, it can be exhaustively divided into a part for each particle.

I shall presently say more about the meaning of 'the complete state of a particle' in terms of regions of space-time, but until assumptions are made about the nature of the $L_n(t)$, separability is almost exclusively a *semantic* claim, and requires only the weak version of Einstein's claim as given above: it is possible to speak meaningfully of the complete state of a single particle in the EPR–Bohm experiment (even if that particle is entangled with some other particle).

There is an obvious objection to taking separability as a condition on models of the EPR–Bohm experiment: namely, that it precludes quantum mechanics itself from being such a model! After all, it is said that quantum mechanics is non-separable. How, then, is it reasonable to require that models of the EPR–Bohm experiment be separable? The answer is that quantum mechanics is *not* non-separable, at least not on my definition. The claim that quantum mechanics is non-separable is based on the fact that some particles — notably those in entagled states — lack wave functions. However, they do not lack *states*. The state of a single particle in the EPR–Bohm experiment is obtained by tracing over the other particle to obtain a so-called 'improper mixture'. It may be claimed that improper mixtures are 'incomplete', i.e., that they do not represent entirely the state of a system — improper mixtures do not encode the correlations that a system might have with other systems. True, but separability allows for such correlations. In other words, quantum mechanics may be seen as a separable theory with correlations between systems, as in the EPR–Bohm experiment. Hence quantum mechanics itself is certainly not excluded from the analysis to follow.

7.2.3 *To what do the complete states refer?*

To what, in general, do the values of the $L_n(t)$ refer? It would be too much to presume that they are 'particles' in the traditional sense, for this assumption would bias the analysis towards interpretations that assume the existence of particles (such as Bohm's theory). On the other hand, separability was motivated by the need to be able to describe a system as localized in a continuously connected region of space-time. If this region is not to be the world line of a particle, then what? A good place to begin is with the

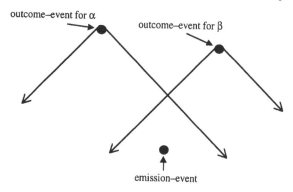

Fig. 7.1. *Space-time representation of an EPR–Bohm experiment.* The space-time diagram here is a 1+1 diagram. The particles fly out in opposite directions, and the outcomes occur at the apex of the light cones, as depicted.

backwards light cone of the event $D_{d,\pm}^{n}(t^{n})$. This region is depicted in figure 7.1 for each outcome.

I shall consider the holist's objection once more. The point may be made that letting the complete states refer to regions of space-time turns separability as given above — which is just the weak form of Einstein's claim — into the stronger version of Einstein's claim. Again, I emphasize that I find this version of the claim very plausible. Nonetheless, it must be admitted that one might accept separability itself while denying that the complete states of individual 'particles' refer to regions of space-time.

Despite this possibility, the backwards light cone of the outcome has much to recommend it as an object for study in the EPR–Bohm experiment. However, clearly it will not do as the region whose state is given by $L_n(t)$, because $L_n(t)$ refers to a single time. The solution I choose is to let each reference frame have its own version of $L_n(t)$, which refers at each time to a space-like slice of the light cone. That is, each reference frame determines a foliation of space-like slices of the backwards light cones, and each observer assigns to each slice a state as given by that observer's version of $L_n(t)$. We need not for the moment make assumptions about the transformation properties of the $L_n(t)$, though, of course, it would be nice if they were covariant.

Note that because the light cones are relativistically invariant, this proposal has at least a chance of yielding a covariant theory. For example, one observer might use $L_n(t)$ to assign states to time-slices as depicted by the dashed lines in figure 7.2, while another would use (a transformed) $L_n'(t')$ to assign states to the time-slices depicted by the dotted lines. Finally, I shall assume that if $t^{\beta} < t^{\alpha}$, then for all times, t, (in a given frame) after t^{β}, $L(t)$ contains

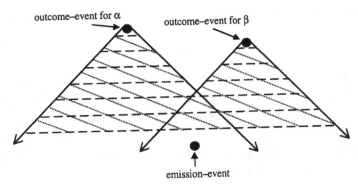

Fig. 7.2. *Time slices assigned states by two different observers.* The dashed lines represent time slices assigned states by an observer whose time axis goes straight up the page. The dotted lines represent time slices assigned states by an observer at motion with respect to the first observer.

the outcome for particle β. (This requirement would be a consequence of a more general proposal to let $L(t)$ cut across the forward and backward light cones of the outcomes, as appropriate, but all I need for later is that $L(t)$ should contain enough information to fix the outcome for β, given that $t > t^\beta$, which is to say that the outcome is *recorded*.[12])

7.2.4 *Two conditions of locality*

Now that we have a reasonable space-time description of the evolution of the pair of particles (I continue to use the word 'particles' for convenience), we may formulate dynamical conditions of locality as conditions of independence of the evolution of the particles. To begin, it seems a reasonable locality principle that the evolution in one of the backwards light cones be independent of the other, except insofar as they intersect:

Independent evolution: For any t, t' such that $t \geq t' \geq t_0$, $\mathsf{E}\left[L_\alpha(t)\big|L_\alpha(t_0), L_\beta(t')\right] = \mathsf{E}[L_\alpha(t)|L_\alpha(t_0)]$, where $\mathsf{E}[\,\cdot\,|\,\cdot\,]$ is a conditional expectation value.

Note in particular that because of λ-determination (which is trivial, given the meaning of the λ adopted here), independent evolution entails in particular that the outcome for one particle does not influence the evolution of the other, given the initial state. In addition to this condition, however, we might like the *outcomes* to be determined solely by the complete state in that region — the natural 'local' version of λ-determination. Then we require:

Local determination: For any n, d, t^n, and d_n^t, and any complete state $L(t^n) = (L_\alpha(t^\alpha), L_\beta(t^\beta))$,

$$p^{L(t^n)}(d_n^t) = p^{L_n(t^n)}(d_n^t).\tag{7.13}$$

One could, of course, define locality conditions other than independent evolution and local determination, but these are sufficient to capture two interesting senses of locality. Together they entail, roughly, that a theory is 'local' if the state of a region of space-time is screened off from all space-like separated events by its backwards light cone. I turn now to see how these conditions are related to determinism.

7.3 Determinism and locality in dynamical models

7.3.1 Deriving determinism from locality

Coupled with the strict correlations, the locality conditions of the previous section entail determinism, in much the same way that factorizability as given in chapter 6 entailed two-time determinism. However, the dynamical case does bring some subtleties, and it is therefore worth going through the details.

Imagine that β is measured at time t^β (in some frame), and that α is to be measured at the same or later time, t^α. The outcome-events are, of course, assumed to be space-like separated. The outcome of the measurement on β is $D_{j,r}^\beta(t^\beta)$, for some $r(=\pm 1)$, and the strict correlations guarantee that the result of the measurement on α will be $D_{i,-r}^\alpha(t^\alpha)$, if (as we may assume) i and j are parallel. In other words, thus far we have:

$$p^\psi(i_\alpha^t = -r \mid j_\beta^t = r) = 1.\tag{7.14}$$

Because of the strict correlation in this case, empirical adequacy will dictate that the conditional probability above is also 1 for every state, λ (more precisely, for measure-one of them, where ρ is the measure):

$$p^{L(t^\alpha)}(i_\alpha^t = -r \mid j_\beta^t = r) = 1.\tag{7.15}$$

Now, because $L(t^\alpha)$ is the *complete* state of *both* regions (and in any case includes the result at β by definition), we may assume that (7.15) entails

$$p^{L(t^\alpha)}(i_\alpha^t = -r) = 1.\tag{7.16}$$

By local determination, (7.16) is

$$p^{L_\alpha(t^\alpha)}(i_\alpha^t = -r) = 1.\tag{7.17}$$

Now consider $L_\alpha(t_0)$. By independent evolution, the evolution of $L_\alpha(t)$ from

t_0 to t^α is independent of the state of β. Therefore, we may ignore β and the measurement on it and write:

$$p^{L_\alpha(t_0)}(i_\alpha^t = -r) = 1. \tag{7.18}$$

In other words, the condition of deterministic results for α follows from local determination, independent evolution, and the strict correlations. The condition of weak deterministic transitions therefore follows as well.

Does the conclusion hold for β also? One might be tempted to say 'yes', on the basis of symmetry (which anyhow would be another assumption), but the situation is not really symmetric. Particle β differs from α in the very important respect of having been measured. Who can say that the outcome for β did not trigger a superluminal signal to α, rendering determined the previously undetermined result of the measurement on α? In other words, the measurement on β could force the state of α to become some state for which the condition of deterministic results holds, but it does not follow that this condition would hold without the occurrence of the outcome for β. Here we must use separability and independent evolution to find that, in fact, it does — the outcome for β can have no affect on the evolution of α. Hence any evolution that $L_\alpha(t)$ might undergo without the outcome for β, it might also undergo with the outcome for β. However, all such evolutions must be such that (7.18) is obeyed. Hence all evolutions of $L_\alpha(t)$ are such as to obey deterministic results. Now, we may adopt a very weak symmetry condition:

Weak symmetry: There is no qualitative difference (as regards deterministic results) between the evolutions of a particle as given by $L_\alpha(t)$ and those given by $L_\beta(t)$ — i.e., the range of possible evolutions is the same for each particle.

Given weak symmetry, the condition of deterministic results does hold for both particles. In other words, the assumptions of weak symmetry, local determination, and independent evolution, together with the strict correlations of the EPR–Bohm experiment, entail deterministic results.

The main idea behind this claim is straightforward. If the two particles are isolated from one another, then in order to maintain their strict correlations, they must have already 'decided' what the results of the measurements will be, no matter when the measurement is made. In fact, though I will not need it, more can be shown. One might hope that the particles could at least decide on *different* (but still strictly correlated) results for measurements at different times, so that deterministic results and weak deterministic transitions both hold, and yet the determined result is different for different times. This strategy cannot work, because one particle cannot tell when the other was measured, and therefore will keep on 'evolving' until it is measured.[13]

Recall from chapter 6 that not only did locality imply determinism, but also determinism implied locality. Does the same hold here? No. The introduction of a fully dynamical model destroys that half of the inference. Indeed, it is easy enough to see that in a deterministic theory, there could be (deterministic) influences travelling faster than light from the outcome-event for one particle to the outcome-event for the other. Hence, unlike the two-time case (where the outcomes were simultaneous), determinism alone is no guarantee of locality. This point will arise again in chapter 9, where I examine the sense in which some interpretations are local or non-local.

7.3.2 Bell's theorem again

The results of this chapter show that if one wants a local theory, one must be willing to countenance deterministic results and weak deterministic transitions. In particular, locality is possible only in a model where the results of the measurements are completely determined ahead of time by the states of the particles. It is usually said that such a picture is dismissed by Bell's theorem. If it were, then there could be no completely local model of the EPR–Bohm experiment, for locality requires such a picture.

However, I have already noted that Bell's theorem does not quite rule out the picture of complete locality and determinism, for it is not only locality, but also λ-independence, that goes into the derivation of Bell's inequality. Recall what λ-independence was:

λ-**independence**: The distribution $\rho(\lambda)$ is independent of the setting of the magnets (or, of the polarizers).

This condition is independent of all of the locality conditions in this chapter, and therefore it appears to be possible that there be a local, deterministic theory that is *not* committed to Bell's inequality. By a 'local theory' I mean, of course, one that obeys the locality conditions laid down in this chapter: independent evolution and local determination. By a 'deterministic theory' I mean one that obeys the forms of determinism that follow from locality: deterministic results and weak deterministic transitions. However unlikely or bizarre we may think it may be that λ-independence should fail, a local, deterministic, model of the EPR–Bohm experiment has not been ruled out.

8

Non-locality and special relativity

Thus far I have been primarily concerned with how an interpretation of quantum mechanics might, or might not, be local. This question is traditionally not sharply distinguished from the question of whether quantum mechanics is consistent with the theory of relativity. However, the two questions are indeed quite distinct, and should be recognized as such explicitly.

8.1 The theory of relativity

8.1.1 What does relativity require?

As a first approximation, we may make the distinction by noting that the requirements of the theory of relativity are themselves unclear. Minimally, relativity seems to require that there be no way to distinguish one reference frame from another — i.e., that there be no experimental procedure that can determine which of two inertial observers is 'really' moving, or more generally that there be no way to discover an observer's absolute velocity.

Most authors are willing to find in (special) relativity a stronger requirement: namely, Lorentz-invariance. They say that relativity requires that *in fact* there is no such thing as absolute velocity. This requirement goes beyond the minimum — it might be that there is an absolute rest frame (so that absolute velocities are given by motion relative to this frame) while there is no way to find it. (As I will discuss in chapter 9, exactly this situation occurs in Bohm's theory.)

Whether one wants to draw this stronger lesson from relativity is largely a matter of taste. However, some may try to argue for the stronger lesson along the following lines. Special relativity is not a phenomenological theory. It is a theory of principle — the entire theory can be derived from just a couple of fundamental principles, one of them being Lorentz-invariance. Indeed, from the beginning, Einstein derived special relativity from just a

163

couple of physical principles. At the least, then, we should be reluctant to give up those principles — the success of the theory of relativity is evidence for their truth.

This argument is indeed powerful, but it is also misleading. What are the arguments for the physical principles in question? How did Einstein really *motivate* these principles? Answer: by considering what sorts of experiments are possible. In particular, Einstein noted that for quite general reasons, it is not possible to find a frame at absolute rest (or, what comes to the same thing in this context, to measure the one-way velocity of light). In other words, the considerations that Einstein used to support his basic principles were ultimately *epistemic*. They were not (directly, at least) evidence against the *existence* of a preferred frame (of absolute rest), but evidence against the existence of any experiment to find such a frame. The theory of special relativity is apparently, then, motivated (in part) by what is experimentally possible, and such a motivation might be taken to support only the minimal requirement given above, rather than the stronger requirement.

Of course, the notion of possibility is rather strong here — Einstein did not argue that the problem of finding an absolute rest frame is a problem of engineering. It is a problem of principle. Hence the move from the minimal to the stronger requirement is not so great a jump as it might at first appear to be. It might be underwritten, for example, by a kind of ontological minimalism: if it cannot be detected, even in principle, then it does not exist. Or, it might be underwritten by some form of Kantianism: what exists (in the physical world) is exactly what can be known to exist. However, we must recognize that these motivations for the stronger requirement of Lorentz-invariance go beyond the content of relativity itself, and even go beyond the immediate motivation for the theory of relativity.

Nonetheless, many are willing to see in relativity even stronger requirements of some form or other. For example, relativity has been taken to forbid cause–effect relations between space-like separated events. It has been taken to forbid superluminal transmission of matter or energy. It has been taken to forbid superluminal transfer of information, or superluminal signalling. And so on. However, *none* of these things is obviously and straightforwardly ruled out by relativity, but I will not undertake a discussion of this point here, for others have already done so at length. In particular, Maudlin has examined in some detail what relativity does and does not require.[1] He concludes (correctly, in my view) that at least some forms of superluminal causation, transmission of matter-energy, and signalling, are permitted by the theory of relativity.

In any case, whatever relativity requires, it is an open question whether a given interpretation of quantum mechanics is compatible with the theory of relativity. Moreover, this question is, in general, quite *difficult* to answer, for a given interpretation, and it is *not* answered by answering the question 'is the interpretation local?' Or anyhow, it is not so answered unless one has already argued that the theory of relativity requires locality, and that claim is by no mean obviously true. I will discuss this point further below.

8.1.2 *Digression:* **The block-universe argument**

The results of chapters 6 and 7 suggest one quick — but ultimately unsatisfactory — account of the relationship between quantum non-locality and special relativity. Those results show that locality requires determinism, and in that case Bell's theorem is avoided only by denying λ-independence. However, some authors have argued that special relativity itself also requires determinism. Hence, at first glance, there is the potential for some exciting connections to be made between local models of the EPR–Bohm experiment and special relativity. Closer examination will make us considerably less excited. In this section, I will examine and reject the argument for determinism in special relativity, the so-called 'block-universe' argument.

Probably most people have something like the following view about the difference between the past and the future: the past is 'fixed', or 'determinate', whereas the future is 'open', is 'yet to occur', and contains a number of possibilities, only one of which will be realized. However, in 1966, Putnam and Rietdjik argued (independently) that the future is 'determined', or, 'fixed', or 'not open'.[2] Their claim is that the theory of special relativity is strictly *incompatible* with the view that the future is open, or indeterminate. The argument advanced in favor of this conclusion is often called the 'block-universe argument', because it is supposed to establish that the universe is 'given once', as a 'block' of space-time. This conception is meant to contrast with a conception of the universe as 'unfolding' over time.

In this subsection I shall do two things. First, I shall attempt to unravel the arguments that Putnam, Rietdjik, and later Maxwell, have made in favor of their claim. I will suggest that there are, in fact, four different arguments being made, of differing plausibility. Second, I shall argue that none of these arguments succeeds in establishing its intended conclusion. More precisely, I shall argue (along lines already suggested by Stein[3]) that the very doctrine that the block-universe argument is meant to overturn — called 'probabilism' by Maxwell — is simply meaningless in a relativistic context. It follows that the conclusion of the block-universe argument — namely, that probabilism is false — is of no interest in a relativistic context.

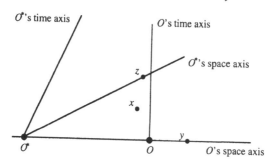

Fig. 8.1. *The geometry of the blockuniverse argument.* The event x is in the future light cone for O, but is below O's space axis.

Finally, I shall suggest that a properly relativistic version of probabilism is unaffected by the block-universe argument.

8.1.2.1 *Four block-universe arguments*

As I said, the block-universe argument is meant to address the doctrine that the future is 'open'. Let us begin in a somewhat naïve way to see how the block-universe argument is supposed to establish the falsity of this doctrine, and then I shall distinguish four different senses in which the future may be said to be 'open'.

The doctrine that the future is open says, *prima facie*, that there is a 'now', and that everything 'to the past of' this now is 'fixed', while everything 'to the future of' this now is 'open'. Or, to put it closer to Putnam's terms, there is a 'now', and everything that is 'now' is 'real', while everything that is 'to the past of' this 'now' is 'unreal, but already become, or fixed' and everything 'to the future of' this 'now' is 'unreal, and not yet become, or open'. For the moment, let us call this doctrine 'probabilism'. Later I shall be more careful to distinguish different senses of probabilism.

The block-universe argument then proceeds as follows. Consider the observer, O, in the space-time diagram, figure 8.1. O's spatial axis divides space-time into a past (everything below the axis), present (everything on the axis) and future (everything above the axis). So far, no problem for probabilism. Now consider a second observer, O^*, who, as far as O is concerned, is 'now', i.e., who lies on O's space axis. O^* is in motion relative to O and therefore has a space axis tilted with respect to O's. Now consider a space-time point, x, which is above O's space axis but below O^*'s. What is O to say about this point? Originally, O said that it was 'open', or 'not yet become', but O must now recognize that for O^*, x *is* 'fixed', or 'already become'.

Now, if special relativity teaches us anything, it teaches us that there are no 'privileged' observers. Hence O must recognize the 'equal authority' of O^* to say what has become and what has not become. To put it differently (and perhaps more convincingly), the principle that there are no privileged observers is meant to imply that for any space-time point, y, and any two observers, O and O^*, if O^* is simultaneous with O and if y is 'fixed' for O then y is 'fixed' for O^*. It follows that the point x considered above is fixed for both O and O^*.

(Note the importance of the clause 'if O^* is simultaneous with O'. If we were to consider just any two observers, the principle would be obviously false. For example, what was definite for Alfred the Great is surely not the same as what is definite for you. On the other hand, it is apparently more plausible to say that what is definite for observers who are simultaneous with you is also definite for you.)

Moreover, by considering observers moving at faster velocities, and at greater distances away from O, we can establish that everything to the future of O's spatial axis is, in fact, definite. (By moving O^* further and further away from O, we sweep out more and more of the future of O with O^*'s spatial axis.)

The geometrical facts underlying this argument are not in dispute, but what exactly are they supposed to entail? I find in the literature four separate arguments, all relying on these geometrical features of Minkowskian space-time, but all differing in their conclusion.

For Putnam, the argument is about 'what is real'. He claims that our everyday notion of time includes the statement that 'All (and only) things that exist *now* are real'.[4] The argument above, then, is supposed to establish that this statement (together with the principle that there are no 'privileged' observers) is incompatible with special relativity. For example, O cannot maintain that only the events on O's own space axis (such as the event y depicted above) are 'real', because O is simultaneous with O^*, for whom events on O^*'s own space axis (such as the event z depicted above) are real. Moreover, as there are no privileged observers, O must recognize that because O^* is simultaneous with O, and because z is real for O^*, z is also real for O.

This argument can be repeated over and over, for different observers O^*, to establish finally that *every* event in space-time is real for O. Putnam maintains that the proper lesson here is to give up the 'everyday' notion of time. For Putnam, then, the block-universe argument establishes that the future (and the past, by the same reasoning) is real right now. The universe does not unfold, one instant at a time; rather, it is given once, as a 'block' of space-time. Let us call this version of the conclusion of the block-universe argument the 'reality of the future' version (though it equally concludes the reality of the past).

Putnam draws the further conclusion that future contingent statements have a truth-value now. According to Putnam, Aristotle argued that contingent statements about the future have no truth-value — Putnam says that Aristotle denied the ' "tenseless" notion of existence'.[5] However, because the future is real even now, future contingent statements must — even now — have a truth-value. Hence, says Putnam, Aristotle was wrong. Let us call this version of the conclusion of the block-universe argument the 'truth of future contingents' version.

Rietdjik draws a slightly different lesson from the same geometrical facts.[6] He supposes that they entail a form of determinism, which he defines as follows (with slight change of notation):

We say that an event x is (pre-)determined if, for any possible observer O, ... who has x in his absolute future [i.e, future light cone], we can think of a possible observer O^* (or: there may exist an observer O^*) who can prove, at a certain moment T, that O could not possibly have influenced event x in an arbitrary way (e.g., have prevented x) at any moment when x still was future, or was present, for O.[7]

The proof of 'determinism' in this sense is supposed by Rietdjik to go as follows. Imagine that O (as depicted in figure 8.1) claims to have some influence over whether an event at x occurs or not. What will O^* be able to say about this claim? Well, for O^*, the event at x is in the past. It has occurred, and its occurrence is a fact of history that cannot be altered. O^* says, therefore, that the event at x is *unalterable as of right now*. However, O is simultaneous with O^* (according to O), so that O must recognize that, in fact, the event at x is unalterable as of right now. In other words, O must recognize that nothing can be done to influence whether the event at x occurs. Let us call this version of the conclusion the 'determinism' version.

Maxwell draws yet a fourth lesson. He makes a distinction between 'ontological probabilism' and 'predictive probabilism', which he defines as follows:

[Ontological probabilism] asserts that the basic laws are probabilistic *and that the future is now in reality open with many ontologically real alternative possibilities* whereas the past is not. ... [Predictive probabilism] asserts that the future, like the past, is now in reality entirely fixed and determined even though the basic laws are probabilistic and not deterministic. According to predictive probabilism, alternative possible futures represent no more than alternative possibilities relative to what can in principle be predicted on the basis of a complete specification of the present, and the basic laws: they are not alternatives *in reality*.[8]

Maxwell 'admits' that his argument (which is again just the block-universe argument) applies only to ontological determinism — not much of an 'admission' in fact, because 'predictive probabilism' itself 'admits' the conclusion that Maxwell hopes to establish, namely, that the future has the same ontological status as the past, that of being 'fixed and determined'.

This conclusion differs from Rietdjik's in that it says nothing about the causal influence of the present on the future. Rietdjik's conclusion was that 'determinism' holds, which means for him, roughly, that actions performed in the present cannot influence whether or not a future event occurs. Maxwell's, on the other hand, makes no reference to the causal relation between actions made in the present and future events. For him, the point seems to be just that the future is ontologically determinate. Let us call Maxwell's version of the conclusion the 'determinateness' version.

8.1.2.2 The arguments rejected

We have, then, four versions of the conclusion of the block-universe argument: the reality of the future version, the truth of future contingents version, the determinism version, and the determinateness version. Do the undisputed geometrical facts cited by the block-universe argument(s) really establish any of these conclusions?

My central claim is that they could not possibly do so, precisely because *none* of these conclusions is stated in language that makes sense in a relativistic context. Moreover, if one does so state them, then they are by no means consequences of the geometrical structure of Minkowskian space-time.

Consider the fact that each of these conclusions makes reference to one or more of the following notions: 'the present', 'the future', 'the past'. (The determinism version is the least obvious culprit here — but even determinism relies on the idea that actions in the 'present' cannot change what would happen in the 'future'.) In order for special relativity to have anything to say about these conclusions, then, it too must have some way of speaking about 'the present', 'the future', and 'the past', but it does not.

In special relativity (as standardly interpreted), there is no such thing as 'the present'. There is just 'the present relative to a given observer'. The notions of an 'absolute present', 'absolute past', and 'absolute future' are simply meaningless in special relativity. This fact is summed up by the phrase 'relativity of simultaneity', which means that two events that are simultaneous for one observer, O, will not be simultaneous for an observer moving relative to O.

It is ironic that the block-universe argument makes *essential* use of the relativity of simultaneity to establish its conclusions, while the conclusions themselves appear to deny the relativity of simultaneity (by supposing that there do, in fact, exist absolute notions of past, present, and future). If I wanted to be uncharitable, I would summarize the logic of these arguments as follows: from the premise of the relativity of simultaneity (which is presumed when we accept special relativity), we infer the truth of a doctrine that presupposes an absolute notion of simultaneity. Being slightly more charitable, I think that the correct diagnosis of these arguments is the following: from the premise of the relativity of simultaneity, these authors infer the falsity of a doctrine (probabilism, non-truth-valuedness of future contingents, and so on) that presupposes an absolute notion of simultaneity; from there, they conclude the *denial* of this doctrine, but still in the context of a supposed absolute simultaneity. For example, Putnam considers the doctrine that all and only those events that are *now* are real. He concludes, on the basis of special relativity, that this doctrine is false, and takes himself to have shown thereby that 'the future' is real. This form of argument is apparently not very compelling.

It is clear what must be done. Before special relativity can have anything to say about the doctrines in question, they must be expressed in a language that is meaningful in a relativistic context. Let us take just a single example, that of Maxwell's argument for the denial of ontological probabilism. To begin, we would have to reformulate the definition of probabilism in a relativistically acceptable way. Perhaps the most obvious way to do so is to make the following definition, which follows as closely as I can the wording in Maxwell's own definition:

Relativistic ontological probabilism: the basic laws are probabilistic and for any given observer, O, the future light cone for O is now in reality open for O, with many ontologically real (for O) alternative possibilities, whereas the past light cone for O is not.

Here I have just relativized the notions in Maxwell's definition that require relativization in order to be meaningful in the context of special relativity.

Can the block-universe argument be used to establish the falsity of relativistic ontological probabilism? Apparently not. The best attempt to bring the geometrical facts cited in the block-universe argument to bear on relativistic ontological probabilism would then proceed along the following lines. Consider an event, x, that is in O's future, i.e., in O's future light cone. Let us grant (for the moment) the following principle:

If x is not 'open' for any observer who is simultaneous with O, then x is not open for O.

Simultaneous according to whom? In fact we need not answer this question, for the principle is insufficient no matter how we answer this further question. What we would need is an observer who is simultaneous (for somebody) with O — and therefore space-like separated from O — but for whom x is in the past light cone rather than in the future light cone. However, there exists no such observer.

My argument is therefore twofold. First, I have argued that the logic of the block-universe argument is flawed, because the attempt is made to use relativity theory itself to address essentially prerelativistic doctrines, and to establish conclusions that are framed in an essentially prerelativistic context. Second, I have argued that if we reformulate Maxwell's definition of ontological probabilism in a relativistically acceptable way, then relativity theory does not show this doctrine to be false, i.e., does not establish Maxwell's conclusion.[9] This argument is easily translated into an argument against the other three conclusions. However, I will now instead consider some objections to my argument.

8.1.2.3 Objections

The points I have just made echo those made in two papers by Stein.[10] The first was a reply to the original arguments of Rietdjik and Putnam, and the latter is aimed more specifically at Maxwell, though applies as well to the earlier arguments. Publications by Rietdjik and Maxwell[11] are curiously devoid of any reference to Stein's first article (which was why Stein had to do it again later!), and so are also devoid of explicit attempts to answer the sorts of argument I have made above.

However, both Maxwell and Putnam did at least consider my second argument, namely, that by taking the past light cone to be 'the past' and the future light cone to be 'the future' (the only truly natural definitions of those terms in special relativity), the conclusion of the block-universe argument is avoided. They each reject this argument, for different reasons. I will consider each of their objections in turn.

Putnam appears to think that this account of past and future is incoherent as an account of which statements have truth-values:

This last move, however, flagrantly violates the idea that there are no Privileged Observers. Why should a statement's having or not having a truth value depend upon the relation of the events referred to in the statement to just one special human being, *me*?[12]

The answer, given my first argument above, should be obvious. In a truly relativistic setting, there is *no such thing* as 'the fact of the matter' about whether a statement has a truth-value. Instead, some statements have a truth-value *for me* and others do not. Therefore, the relativized doctrine says *not* that whether a statement *absolutely* has a truth-value depends on just one observer, *me*, but rather that whether a statement has a truth value *for me* depends on just one observer, *me*. And surely that conclusion is entirely satisfactory.

We must be careful, of course, to keep in mind that all of this discussion is in the context of Aristotle's doctrine that future contingent statements lack truth-values. It may, of course, be that the idea that statements about the future lack truth values is already unacceptable. However, if it is not, or not obviously so (as Putnam's discussion must, and does, assume), then it is no good to object to the relativized version by claiming that the notion that whether a statement has a truth-value must be relativized to observers is unacceptable. In a relativistic context, if we are to discuss Aristotle's doctrine at all, then we *must* accept the *prima facie* possibility that whether some statements have a truth-value will be relative to observers, because Aristotle's doctrine is that whether some statements have a truth-value is relative to a time, and, in relativity theory, time is relative to observers. Without relativizing Aristotle's doctrine to observers, therefore, relativity theory would have no way to discuss it in the first place.

Maxwell has a different objection. He considers the relativized version of ontological probabilism that I gave earlier, and notes that on this view, there will be events simultaneous with me that have the ontological status of 'future' events — they are 'open', or indeterminate'. He concludes:

But this suggestion faces the fatal objection that it postulates not just future alternative *possibilities*, but present alternative *actualities* — a full-fledged multi-universe view. ... This [suggestion] thus commits us to the view that whenever anything probabilistic occurs, there being N equally probable outcomes, three-dimensional space splits up into N distinct three-dimensional spaces, each space containing one of the N outcomes.[13]

Maxwell considers this consequence 'too grotesquely *ad hoc* to be taken seriously'.[14]

However, the flaw in Maxwell's reasoning should by now be painfully clear. He has assumed, again, that there is an 'absolute now' in special relativity, and has failed to appreciate the distinction between an event's being 'indeterminate' and its being 'indeterminate for me'. This assumption is clear in his statement that 'three-dimensional space splits up'. Maxwell's reference to three-dimensional space reveals that he is imagining an 'absolute now' (which defines a hypersurface of simultaneity, and thereby a three-dimensional space) that is 'the time' at which the three-dimensional space 'splits up'. There is no such thing.

Moreover, the fact that Maxwell considers there to be any need at all for a splitting suggests that he is identifying 'what is determinate for me' with 'what is determinate, *simpliciter*'. Having made that identification, it seems necessary to say that if something is 'indeterminate for me', then this indeterminacy must be reflected, somehow, in 'reality', and apparently the only way Maxwell can imagine this mirroring of 'indeterminacy for me' by reality is to have reality split into parts, each one representing one of the possibilities left open by the indeterminacy. However, there is no warrant for the identification of 'indeterminate for me' with 'indeterminate in reality' in the first place, and indeed this identification seems to rely again on the illegitimate notion of an absolute now. For what relativistic ontological probabilism says is that 'determinateness' is relative to observers. Hence, although it makes sense to say that what occurs at a given event is, relative to me-now, indeterminate, it does not make sense to conclude that reality itself is therefore indeterminate right now, for there is no 'right now' in reality. In other words, according to relativistic ontological probabilism, the notion of 'indeterminateness' makes essential reference to the 'now' for a given observer, and there is therefore no way to translate that notion into any statement about 'reality itself', in which there is no notion of 'now'.

There are, then, two mistakes in Maxwell's objection. First, he has apparently introduced a notion of absolute time in his claim that the universe must (according to the relativized version of ontological probabilism) split 'at a time', or more precisely, that certain three-dimensional spaces must split. Second, he has apparently identified 'indeterminate for me' with 'indeterminate *simpliciter*'. However, it is exactly the latter notion that the relativized version of ontological probabilism is meant to avoid, for it is exactly this notion that finds no room in relativity theory (because it requires an absolute now).

What looked like an enticing opportunity to draw a strong connection between (the necessary determinism of) special relativity and (the necessary determinism of) local models of the EPR–Bohm experiment has, alas, turned out to be a Siren's song. As far as I can tell, there is no generic connection between the requirements of special relativity and the requirement of local models of the EPR–Bohm experiment. Instead, one must take each model on its own and discover its relation to special relativity. Doing so obviously requires, first of all, that the model itself be sufficiently well developed. In particular, a model must have a complete dynamics — we have already seen that non-dynamical models are too simplistic — and it must be somehow susceptible to a space-time description. Claims made on behalf of models not meeting these conditions should be approached very cautiously, and with considerable skepticism.

8.2 Probabilistic locality and metaphysical locality

8.2.1 Probabilistic locality

8.2.1.1 Locality as a screening-off condition

In the previous two chapters, we encountered several 'locality' conditions, all meant to capture the idea that the particles in the EPR–Bohm experiment are independent (chapter 6) or evolve independently (chapter 7) of one another, and that the outcome-events in the EPR–Bohm experiment are correlated only indirectly, through correlations between the particles. These conditions may be called 'probabilistic' because the notion of independence that they employ is probabilistic independence.

More precisely, they employ a notion of 'screening off'. We may say that for any events, A, B, and C, if $p(A|B, C) = p(A|C)$ and $p(B|A, C) = p(B|C)$, then C 'screens off' A and B from each other. This condition is due to Reichenbach.[15] Philosophical discussions of screening–off — and in particular the relationship between screening–off and causality — have become somewhat esoteric, and my discussion here is, by present standards, simplistic. However, it is sufficient for now to illustrate the main point, which is that the locality conditions that I have thus far considered are couched purely in terms of probabilities.

There is a somewhat subtle difference between the conditions of chapters 6 and 7 and traditional screening-off conditions. The latter are given in terms of conditional probabilities, as above, but the locality conditions given here do not *condition* on the complete state of the system. In the terms given above, they are written $p^C(A|B) = p^C(A)$ and $p^C(B|A) = p^C(B)$.

I will focus in the next subsection on just two conditions from chapters 6 and 7, Bell-factorizability and independent evolution. The first, recall, is that for any λ, i, j, i_α, and j_β, $p^\lambda(i_\alpha, j_\beta) = p^\lambda(i_\alpha) \times p^\lambda(j_\beta)$. This condition is clearly a screening-off condition of the sort I have mentioned, for it is a general theorem of classical probability theory that $p(A, B) = p(A)p(B)$ if and only if $p(A|B) = p(A)$ and $p(B|A) = p(B)$.

Independent evolution is likewise a kind of screening-off condition. Recall what it was: for any t, t' such that $t \geq t' \geq t_0$, $E[L_\alpha(t)|L_\alpha(t_0), L_\beta(t')] = E[L_\alpha(t)|L_\alpha(t_0)]$, where $E[\cdot | \cdot]$ is a conditional expectation value. In this case, of course, the screening-off condition involves not probabilities, but expectation values.

8.2.1.2 Is probabilistic locality entailed by relativity?

Does relativity have anything to say about these locality conditions? I have already argued that it does not — in particular, relativity is compatible with both determinism and indeterminism. In the previous section, I made this argument negatively, arguing that the block-universe argument fails. In this subsection, I will make a positive argument, showing precisely how a properly relativistic version of indeterminism is compatible with both quantum mechanics and relativity theory. Of course, a model of this sort cannot be local.

As in chapter 7, consider two foliations of space-time, hyperplanes of which are labelled t and t' respectively. One frame uses the stochastic process $L(t)$ to describe the complete state of the particles in the EPR–Bohm experiment, and the other uses $L'(t')$. Hence each frame has different probability measures, $p^{L(t)}$ and $p^{L'(t')}$, to describe the experiment. For any i and j, any results i_α^t and j_β^t, and any times of measurement t^α and t^β, let the probabilities in the first (unprimed) frame be given by

$$\left. \begin{aligned} p^{L(t_0)}(i_\alpha^t) &= p^{L(t_0)}(j_\beta^t) = 1/2, \\ p^{L(t<t^\alpha)}(i_\alpha^t) &= 1/2, \\ p^{L(t<t^\beta)}(j_\beta^t) &= |\langle j_\beta | L_\alpha(t^\alpha)\rangle|^2, \end{aligned} \right\} \tag{8.1}$$

where $|L_\alpha(t^\alpha)\rangle$ represents the spin of α as contained in the complete state $L_\alpha(t^\alpha)$. (Recall the assumption of λ-determination in chapter 7.) In the primed frame, the probabilities are

$$\left. \begin{aligned} p^{L'(t'_0)}(i_\alpha^{t'}) &= p^{L'(t'_0)}(j_\beta^{t'}) = 1/2, \\ p^{L'(t'<t'^\beta)}(j_\beta^{t'}) &= 1/2, \\ p^{L'(t'<t'^\alpha)}(j_\alpha^{t'}) &= |\langle i_\alpha | L_\beta(t'^\beta)\rangle|^2, \end{aligned} \right\} \tag{8.2}$$

with a similar convention defining $|L_\beta(t'^\beta)\rangle$.

Under a suitable completion, this model is exactly the orthodox interpretation with the projection postulate, where in the unprimed frame, the measurement on α occurs before the measurement on β, while in the primed frame the order is reversed. However, the point here is just to illustrate that once we have relativized probability measures to frames (which one *must* do in any case!), indeterministic models can be found that are both consistent with relativity and quantum mechanics. Consistency of the model above (or rather, a reasonable extension of it to a complete model — not all probabilities are given above) with standard quantum mechanics is already evident because it is just the orthodox interpretation with the projection postulate. The model is also consistent with the theory of relativity, at least in its minimal form. The quantum-mechanical no-signalling theorem guarantees that no experiment can be done to distinguish one frame from another. The proof of this theorem is simple.[16] It proceeds by averaging probabilities on one side over the possible results at the other side, and in both frames given above, the result is 1/2. Hence an observer at one measurement-event (who is presumed to be ignorant of the result at the other side) cannot distinguish one frame from another, because every observer will *always* see probabilities of 1/2.

We may recap the logic of the argument thus far as follows. Special relativity itself is compatible with both determinism and indeterminism. Compatibility with indeterminism follows from the model just given. However, the results of chapter 6 and 7 show that indeterminism implies non-locality. It follows immediately that special relativity *is* compatible with at least the forms of non-locality of chapters 6 and 7. Similarly, there exist deterministic models (*viz.*, Bohm's theory). I will discuss Bohm's theory further in the next chapter, but we may note immediately that for some suitable choice of λ, it must be Bell-factorizable — two-time determinism and Bell-factorizability are *equivalent*. However, whether independent evolution also holds — and in the end it is this form of locality that is more appropriate in a fully dynamical theory — is a difficult issue in Bohm's theory. Recall that while the dynamical locality conditions of chapter 7 entail determinism, the reverse implication does not hold. One may therefore evaluate the status of independent evolution only in the context of a given model. In the next chapter, I will do so for Bohm's theory.

8.2.2 Metaphysical locality

8.2.2.1 Is metaphysical locality entailed by relativity?

As I have emphasized, the locality conditions of chapters 6 and 7 are couched in terms of probabilities. However, one may be ultimately interested not so much in whether certain statements made in terms of probability theory are true, but rather in what the world is like. In particular, one may be interested in whether there are *connections* of some sort between the two wings of the EPR–Bohm experiment. Is there a causal connection between them? Is there a flow of information between them? I have already said that I will not undertake a detailed discussion of this point — Maudlin has argued quite convincingly that relativity is *not* incompatible with various types of 'non-local connection'.[17]

I shall not recount the details of Maudlin's argument. In any case, it should not be surprising that such an argument could be made. The theory of relativity is a theory about space-time. The *objects* of the theory, insofar as there *are* objects, are events, points of space-time. In special relativity, the addition of the metric completes the theory. In general relativity one adds the Einstein tensor and the stress-energy tensor. Admittedly, the latter represents the 'matter' in the universe, but nonetheless, general relativity is not a theory about how bits of matter interact with one another.

Indeed, the general reason that Maudlin's arguments work is that neither special nor general relativity is about how matter interacts with matter and hence neither puts much restriction on *how* matter can interact with matter. Of course, one can introduce forces into the theory — one can define force fields, or potential fields, of various kinds, and in this case, the only restriction from relativity is that the symmetries of space-time be respected at least observationally. However, there is no apparent reason that these fields cannot be 'non-local'. The restriction of obeying the symmetries of space-time (whether the Minkowskian space-time of special relativity or the more general space-time of general relativity) leaves plenty of opportunity for 'non-local' action. One easy example is afforded by the invariant hyperbolae in Minkowskian space-time. A causal process, or information, or the like, could propagate along an invariant hyperbola. Such propagation would certainly be superluminal, but also invariant. (Maudlin uses this same example.)

Moreover, there is a more general reason to wonder whether relativity can teach us anything about any form of metaphysical locality. I have already suggested that relativity is best considered a kind of phenomenological theory — it is about what we, as experimenters, can do, or at least about how the structure of space-time restricts what we can know via experiment. If this view of relativity is correct, then relativity is simply not the sort of theory that *could* tell us about things like causal connections and transfer of information. It is quite widely accepted amongst philosophers of physics that the existence of things like causal connections and transfer of information is underdetermined by theory. In the case of relativity, this claim seems more plausible than usual — for relativity theory, in fact, does *not* seem to be about these things.

What? Do we not hear talk, in relativity theory, about 'causal connectibility' and the like? Is relativity not exactly about causal relations? In the present context, this way of speaking about the content of relativity theory is misleading. The claim that only time-like related events are 'causally connectible' is better read as the claim that particles cannot move faster than light, i.e., that they must move along time-like world lines. Only if causal connections must be mediated by particles may we conclude that causal connections can exist only between time-like separated events. Even if we *do* wish to suppose that causal connections are always mediated by particles (quite a substantive assumption — and no part of relativity theory *per se*!), in fact relativity theory is compatible with particles moving at superluminal velocities,[18] as is well known. Relativity theory simply does not rule out superluminal causes.

Although I have focused on non-local causes, similar arguments can be made for other forms of metaphysical non-locality. Relativity is a theory about how things can *move* — it just does not tell us about what *kinds* of things there are in the world, nor how they do, or do not, interact.

8.2.2.2 *Is metaphysical locality entailed by probabilistic locality?*

I have already mentioned that the locality conditions of chapters 6 and 7 are essentially screening-off conditions. Reichenbach introduced such conditions to get at the notion of causal connection. In particular, if two events, *A* and *B*, are correlated, then Reichenbach says that the correlation is the result of a common cause, *C*, just in case *C* screens *A* and *B* off from each other.

The debate over whether causation can be defined in terms of probabilities is difficult, and I shall not take it up here, except to recall that plenty of philosophers have serious doubts about whether there is any significant relationship between conditional probabilities and causality. Moreover, even if such a relationship does exist, simple screening-off conditions like the locality conditions used here may not be sufficient to capture the presence (or absence) of causal connections in the EPR–Bohm experiment. Careful argument, most likely in the context of a specific model of the experiment, would seem to be required to make the case that violation of locality conditions indicates the presence of a causal connection between the wings of the experiment.[19] Indeed, in the end, the lesson of this chapter is that judgments about causality, locality, determinism, and the relations among them, cannot be made at highly general levels. Such judgments must instead be made on a case-by-case basis, in the context of a complete (including dynamically complete) model of the EPR–Bohm experiment. As nice as it would be to have them, highly general results do not seem to be forthcoming.

9

Probability and non-locality

9.1 Review and preview

In chapters 6 and 7, I described some links between locality and the treatment of probabilities in models of the EPR–Bohm experiment. From chapter 6, the main lesson was that, given the strict correlations of quantum mechanics, factorizability of the two-time probabilities is equivalent to two-time determinism. I also noted there that an adequate factorizable theory — one that avoids Bell's theorem — is possible only if λ-independence fails. In chapter 7, I argued that weak symmetry, local determination, and independent evolution together entail deterministic results and weak deterministic transitions.

In chapter 8, I discussed the relationships among the locality conditions of chapter 6 and 7, Lorentz-invariance, and 'metaphysical' locality, especially local causality. I have by no means tried to give general answers to the questions raised there. Indeed, part of my thesis is that there is no general answer to be had. Nonetheless, one *can* use the results of chapters 6 and 7 to investigate the status of various interpretations. Most obviously, they can be used to evaluate whether a given theory is local: if a theory can be shown to violate one or more of the types of determinism entailed by the locality conditions of chapters 6 and 7, then that theory must be non-local in some sense. The particular form that the non-locality takes — and in some cases the form it takes is connected with the determinism or indeterminism of the theory — might help us to see whether the theory is, for example, causally local.

In this chapter, I evaluate some of the interpretations discussed in chapters 2–5 in this way. As my aim here is not to pronounce the final word on any of the interpretations of chapters 2–5, I will not consider all of them. Instead, I will choose some representatives and begin to discuss them (some at greater

179

length than others) in hopes of pointing to one useful way to investigate the status of locality and Lorentz-invariance in interpretations of quantum mechanics.

Of course, we already know that the indeterministic theories *must* be non-local in some way. The status of the deterministic theories is less clear at first glance. What I hope to do in this chapter is to get some grasp of the *sense* in which some of the indeterministic theories are non-local, and to begin an investigation of the locality (or non-locality) of the deterministic theories.

My conclusion will be that none of the deterministic theories is obviously non-local. I shall concentrate on Bohm's theory, and I shall try to show how the determinism of Bohm's theory drives this conclusion. I shall also discuss the status of λ-independence in Bohm's theory, showing why it is violated. In the end, however, my suggestion will be not so much that Bohm's theory may be local, but that the very concept of (metaphysical) locality may simply be inappropriate to completely determinstic theories such as Bohm's theory.

What should we learn from these results? In the last section of this chapter I address this question. The obvious lesson is that if one wants a local theory, then deterministic theories are the only possibility among those discussed in chapters 2–5. However, there is the deeper question of what criteria one is going to use to select a theory. I consider this question briefly, arguing that at least on the criterion of 'conceptual unity' Bohm's theory does not lose out.

A word of caution: as has been the case throughout this book, there is no question of addressing every issue that I raise, nor even of addressing any of them completely. My aim, as always, is more to make suggestions about how the questions might best be addressed.

9.2 Orthodox interpretations

9.2.1 Non-locality and the projection postulate

Standard quantum mechanics with the projection postulate is easily seen to be non-local in the ways described. For one thing, it is evidently stochastic in all of the senses discussed in chapters 6 and 7, and from that fact alone we know that it must at least violate the locality conditions of chapters 6 and 7. This point can be seen directly as well. For consider what happens when the two measurements in the EPR–Bohm experiment occur at space-like separation, one just before the other (in some frame of reference). The first measurement (on α, let us say) instantaneously collapses the state of α (in relativistic quantum mechanics too!), thereby collapsing the state of β, because the two states are correlated. The fact that the collapse occurs instantaneously also reveals that the projection postulate is not relativistic — it must select some hypersurface to define the instant at which the collapse occurs.

There have been attempts to describe the collapse of the wave function relativistically, but there are good reasons for thinking that these attempts are not satisfactory. As Maudlin notes,[1] collapse along the forward light cone of an event is unsatisfactory, because it fails to account for the correlations at space-like separation, in the EPR–Bohm experiment, for example. The other obvious possibility for a relativistically invariant collapse is collapse along the backwards light cone.[2] However, apart from perhaps allowing backwards causation (an oddity in itself), such a theory is hardly a 'collapse' theory (again, Maudlin's point), because in this case the collapse 'occurs' *before* the event that caused it. Hence the statevector was 'collapsed' all along.

Of course, the formal violation of locality does not mean that standard quantum mechanics is non-local in the metaphysical senses described in chapter 8. It might be, for example, that collapse is non-causal, or does not permit signalling, and so on. Such questions must be addressed one at a time.

I shall not undertake to do so here, however. Others have addressed at least some of them.[3] I pause only to note that some authors have noted that standard quantum mechanics appears to violate Jarrett's outcome independence, but not his parameter independence, and they conclude that the violation of locality is explained by a kind of quantum holism. I have already indicated my dissatisfaction with holism, and I will address it again later in this chapter. One of the points that I will make there — and it is important enough to say here too — is that even if we accept holism, it gets us nowhere with the problem of reconciling standard quantum mechanics with the theory of relativity.

9.2.2 Non-locality in CSL

9.2.2.1 The consequence of indeterminism

Because of the stochastic field that is introduced into the Schrödinger equation, CSL is indeterministic in all of the senses discussed in chapters 6 and 7, though there is a slight distinction between two interpretations of the stochastic field. In one, it represents a real physical process acting on the wave function. For example, it has been suggested that this field could represent stochastic (because unknown) effects of background radiation, or of gravity.[4] In the other interpretation, the stochastic field represents a fundamental chanciness in the evolution of the wave function of any system. In the first interpretation, the complete physical state of a system at a time is probably best taken to be the wave function at that time plus the state of the stochastic field at that time. In the second interpretation, the complete state is better taken to be just the wave function, for then the stochastic field represents nothing physical, but is just the means by which fundamental randomness is encoded in the theory.

In both versions of CSL, the conditions of deterministic transitions and deterministic results fail. The former fails because the evolution of the wave function (or, the wave function plus the stochastic field) is a stochastic process. The latter fails for the following reason. The equivalence classes in CSL at the time of measurement are the reduced state vectors, but the probability that the complete state will be in one of these classes given the complete state at any earlier time is not 0 or 1. Therefore, no matter how the equivalence classes for times prior to measurement are defined, the condition of weak deterministic transitions fails. The condition of deterministic results fails for the same reason: given the wave function and the state of the stochastic field at a time $t < t^n$, it is impossible to say how the stochastic field will evolve to collapse the wave function.

What are the consequences of this indeterminism? Using the entailments summarized in section 9.1, we find that one or more of weak symmetry, local determination, independent evolution, or strict correlations must be false. Which of these conditions does CSL violate? Immediately we can rule out weak symmetry and strict correlations. Weak symmetry holds, because the evolutions that CSL allows for one particle, it allows for the other. The strict correlations hold as well, because, as discussed in chapter 2, CSL reproduces the probabilities of quantum mechanics.[5] Therefore, one or both of local determination and independent evolution must fail.

The condition of local determination holds, for the occurrence of an event, $D^n_{d,\pm}(t^n)$, is completely decided by the CSL wave function at t^n in the region where $D^n_{d,\pm}(t^n)$ occurs. However (and therefore), the condition of independent evolution fails. This failure comes *not* from the stochastic field, but from the quantum-mechanical entanglements. One might suspect that the stochastic field is directly responsible for the failure of independent evolution, because it looks as though the fluctuations in this field 'conspire' to insure that one and only one term in the superposed state grows, while the others decay. However, there is no non-local conspiracy here. The effect is purely classical: the probability of a growth in more than one region is negligibly small. Instead, the failure of independent evolution comes from the quantum-mechanical entanglement of the states of the two particles. If α experiences a reduction, β will be carried along (through its entanglement with α) and experience a corresponding reduction, which cannot be anticipated given only the wave function in the region around β. To put it differently, the non-locality implied by the projection postulate is inherited by CSL — the fact that CSL *models* collapse rather than postulates it does not get rid of the non-locality implied by collapse.

9.2.2.2 *Relativistic CSL and elements of physical reality*

The CSL modification of the Schrödinger equation is therefore non-local: it does not obey independent evolution. However, perhaps this result is no surprise, coming as it does from a non-relativistic equation. Well, we have already seen that this move fails in general — relativistic quantum mechanics with the projection postulate is non-local (and, in some sense at least, non-Lorentz-invariant!). Nonetheless, it will be helpful to see that relativistic CSL is also non-local.

The attempts to find a relativistic field-theoretic version of CSL have met with moderate success.[6] The one obvious obstacle, which is also an obstacle to full Galilean invariance, is the fact that reduction is not a reversible process. That is, given a reduced wave function, it is impossible to tell from what wave function it evolved. This fact is evident in standard quantum mechanics with the projection postulate, where projection is complete — no tails are left afterwards — but it appears to be true in CSL as well. Although there are, after a reduction, tiny 'remnants' of the original wave function, it appears that these remnants cannot be used to reconstruct the original wave function. (See the references in note 6 for a discussion.) Hence even non-relativistic CSL does not enjoy complete Galilean invariance; in particular, it is not time-reversible, and CSL must settle for a semi-group invariance.[7]

In relativistic CSL this problem comes back in spades. Consider, for example, what would happen for a CSL modification of the Dirac equation. In the usual Dirac equation, the statevector transforms according to the Lorentz-transformation. However, in every Lorentz-transformation from one hyperplane, H, to another, H', part of H' will lie to the past of H. CSL cannot provide a prescription for getting from H to anything in its past, and therefore cannot provide the necessary Lorentz-transformation from H to H'.

To get around this problem, advocates of CSL have used instead the Tomonaga–Schwinger equation, in which the statevector is assigned not to a hyperplane, but to a hypersurface, σ, and is therefore written $|\psi(\sigma)\rangle$. Then they consider only transformations from one hypersurface, σ, to a second

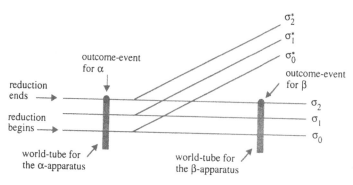

Fig. 9.1. *Hyperplane dependence of local properties in the CSL version of the EPR–Bohm experiment. The properties assigned to the apparatus for α depend on which hyperplane one considers.*

hypersurface, σ', that is at every point coincident with σ or to the future of σ.

There are now two questions to ask about this theory. First, is it invariant under tranformations from one hypersurface to another? Second, is it a plausible physical theory?

The answer to the first question is 'partly'. Advocates of CSL have themselves pointed out that Lorentz-invariance holds *only* for the statistical predictions made by various observers: given a state $|\psi(\sigma)\rangle$, the statistical predictions for various reduced states on the hypersurface $|\psi(\sigma')\rangle$ (where σ' is everywhere coincident with or to the future of σ) are the same for every observer. However, Lorentz-invariance does not hold at the level of the individual system: the properties assigned to a given space-time region depend on the hyperplane one uses to assign those properties.

This fact leads to the second question — Is relativistic CSL a plausible theory? — to which the answer is 'maybe not'. To see the problem consider again the EPR–Bohm experiment, as depicted in figure 9.1. As Ghirardi and Pearle do,[8] consider the hypersurfaces σ_0, σ_1, σ_2, σ_0^*, σ_1^*, and σ_2^*. On the hypersurface σ_0, neither particle has yet reached its apparatus, and the apparatuses are therefore in a 'ready' state. On σ_1, the reduction is in progress, and each apparatus is in a superposed state. By σ_2, the reduction is over, and the apparatuses indicate the results.

Suppose that the actual results in a given run of the experiment are $D^\alpha_{i,+1}$ and $D^\beta_{j,-1}$. Now consider σ_0^*. On σ_0^* the apparatus for β shows the result -1, while α has not yet reached its apparatus. Hence the apparatus for α, which is not yet entangled with anything, is in the ready state. By contrast, on σ_1^* α has reached its apparatus, which is therefore entangled with the other apparatus. The reduction of the state of the other apparatus therefore entails the reduction of the state of the apparatus for α, which therefore shows the result $+1$. On σ_2^*, each apparatus shows a result.

Thus the property assigned to the apparatus for α in the space-time region occupied by its world-tube between the hypersurfaces σ_0 and σ_2 depends on which hypersurface one uses to assign the properties. From σ_1, the state of the apparatus is still unreduced, while from σ_1^* it is reduced. The result is a form of non-invariance: two observers assign different states to the same region of space-time. Moreover, assuming that 'is reduced' is a Lorentz-invariant property (as it seems to be), then apparently these observers *should* agree about the state of this region (at least as far as it concerns reduction). However, they do not.

To resolve this problem, Ghirardi and Pearle have suggested the following criteria for the attribution of properties to physical systems:

Consider a local observable A with compact support α on a space-like surface and one of its eigenvalues, say a. We state that the physical system has the objective property $A = a$ iff the probability of getting the result a, as a consequence of a system-apparatus interaction, is extremely close to one on *any* space-like surface containing α.[9]

This criterion legislates away the troublesome cases — one attributes 'objective' properties to a system only when doing so is unproblematic. In particular, an observer, O, whose hyperplane of simultaneity passes through the world-tube for α may not attribute an 'objective' property to α if another observer, whose hyperplane passes through α's world-tube at the same event and who is on O's hyperplane, would disagree about whether α possessed this property.

Figure 9.2 depicts the region of space-time in which the apparatus for α has no 'objective' properties — call this region 'the proscribed region' for that apparatus. All of the 'non-locality' in relativistic CSL involves 'properties' that systems 'possess' only in their proscribed regions. As the advocates of CSL say themselves, the rule for ascribing properties to systems guarantees that 'no objective local property... can emerge as a consequence of a measurement occurring in a space-like separated region'.[10]

There are three problems with this account. First, to say that the state of the apparatus for α changes as a result of the measurement on β, but that no non-locality is involved because this state does not represent a physically objective property, is a cheat. It is hardly convincing to 'fix' a non-local theory by postulating that the states implicated in non-locality are not physically objective. Anyhow, even if one swallows this move, is

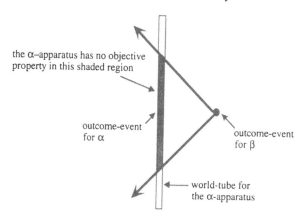

the α–apparatus has no objective
property in this shaded region

outcome-event
for α

outcome-event
for β

world-tube for
the α-apparatus

Fig. 9.2. *Proscribed region in the CSL version of the EPR–Bohm experiment.* The apparatus for α has an objective indicator-state only outside of the shaded region.

'non-objective' non-locality any better than 'objective' non-locality? It is not clear to me that it is.

Second, and related to the first problem, the way in which the advocates of CSL have 'restored objectivity' is not very satisfying. Originally, the CSL wave function was meant to describe 'objective reality'. The fact that different observers will disagree about what the 'objective reality' is, if they use the CSL wave function, was taken to be a blow against the objectivity of the theory. This feature was 'restored' by postulating that what observers *take* to be 'objectively real' is not so, *if* other observers (of the relevant sort) take something different to be 'objectively real'. Although I am not in general of the view that CSL is *ad hoc*, this move certainly is.

Third, it seems likely that, on this account, *very few* of the properties that we suppose to be 'objective' will turn out to be so. There are many ways to make this point. For one, consider the situation depicted in figure 9.3. There, a third particle, γ, is entangled with β, which is entangled with α. (There is certainly nothing unusual about this situation — indeed, almost any time that particles interact, they become entangled with one another.) This extra entanglement extends the proscribed region for α. Such extensions could occur quite generically, and could be quite signficant, if γ travels far from β before it experiences a reduction, or if γ was far away from α when it became entangled with β in the first place.

The problem with such extended proscribed regions is that then almost nothing that we take to be 'objectively possessed' is so. As long as we were asked to believe in 'non-objective properties that are, nonetheless, possessed with probabiltiy 1' only in cases where they very quickly become objective

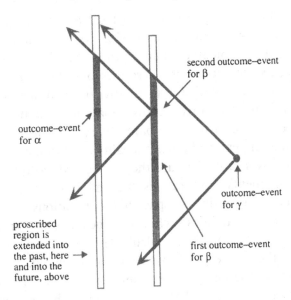

second outcome–event
for β

outcome–event
for α

outcome–event
for γ

proscribed
region is
extended into
the past, here →
and into the
future, above

first outcome–event
for β

Fig. 9.3. *Objectivity in CSL?* Uncontrollable entanglements might make proscribed regions the rule rather than the exception.

(as in standard versions of the EPR–Bohm experiment), we might have been able to accept them. However now, it seems, they are quite generic, a situation that is somewhat more difficult to accept.

Hence the situation with CSL, in my view, is this: CSL is non-local by violating independent evolution; moreover, CSL has serious trouble in the relativistic domain, because its main attraction — its ability to assign an objective physical state based on the quantum wave function — has thus far been saved only by an unacceptably *ad hoc* postulate. It appears that CSL can be 'relativistic' only in the weakest sense. Why, then, play with objectivity? Why not, instead, adopt a preferred frame, and preferred foliation of space-time, in which reduction 'really' occurs. Experimental consistency with special relativity will remain,[11] but without the philosophical extravagence of redefining 'objectivity'.

9.3 No-collapse interpretations

9.3.1 The bare theory: Locality at last

No-collapse interpretations appear, at first glance, to have the means to avoid violations of locality (and perhaps also Lorentz-invariance), because they at least deny collapse, which was the culprit in the orthodox interpretations. However, merely to deny collapse is insufficient, as we will see. Nonetheless, the bare theory does manage to be both local and relativistic. It is local because the quantum-mechanical state of a system is its complete state, and (assuming that the potentials that appear in Schrödinger's equation are local) Schrödinger's equation alone describes only local interactions. It is relativistic because the statevector in relativistic quantum mechanics is a genuine 4-vector; it has the correct transformation properties.

There is a deeper reason that the bare theory is local, which is that the very quantities that appear in the Bell's locality condition do not correspond to anything in the world, in the bare theory. According to the bare theory, recall, the measurements have no results, and if there are no outcome-events, then there can hardly be space-like correlations among outcome-events. Conditional probabilities such as the ones that appear in the locality conditions of chapter 6 and 7 are calculable in the bare theory, but they mean nothing — they correspond to nothing in the world.

I will not consider the many minds interpretation here, but it is local for essentially the same reason, and this in *spite* of its being stochastic. That fact does not constitute a counterexample to my arguments in chapters 6 and 7 that locality implies determinism, because those arguments occur in a framework that assumes that there are outcomes of the measurements in the first place.

I have already given my reasons for being dissatisfied with the bare theory and the many minds interpretation — if the price of locality is to give up on the idea that measurement have outcomes at all, then I am willing to forget about locality.

9.3.2 *Modal interpretations*

Some advocates of modal interpretations have written quite a lot about the status of locality in their interpretation, and especially in the context of the EPR–Bohm experiment. Much of what they have written is an attempt to spell out the details of a 'holistic' account of the correlations, in which one does not end up saying that some localized system 'is the sole cause of' the properties of another at space-like separation, but rather that the correlations are the effect of a single 'holistic' property that somehow attaches to both systems.[12] For reasons partly given in chapter 7, I do not find such an account of the EPR correlations particularly satisfying. However, I shall not have much to say about it here. (Later in this section I briefly address the question of whether the holistic account brings these interpretations any advantage. My answer is 'no'.) Instead, I stick more or less to the questions already addressed for CSL and the methods already used to answer them.

9.3.2.1 The Copenhagen variant

The Copenhagen variant of the modal interpretation is indeterministic. Or rather, if it were endowed with any dynamics, that dynamics would (necessarily) be stochastic. It would violate the conditions of deterministic results, deterministic transitions, and weak deterministic transitions. As for CSL, we may conclude that one or both of the conditions of local determination, and independent evolution must fail. As for CSL, the culprit is independent evolution.

The complete state of a system in the Copenhagen variant, as in all the modal interpretations, is given by either its physical state alone or by its physical state along with its theoretical state. In either case, the following discussion illustrates the failure of independent evolution.

One might hope to catch the Copenhagen variant in a violation of independent evolution in the usual EPR–Bohm experiment, but to do so is more difficult than it might appear. To see why, consider a 'measurement' of whether a system is in the singlet state. As a result of this measurement, assuming that it is non-disturbing,[13] the Copenhagen variant attributes a state to both the apparatus and the pair of particles. In this case, there is only one possible 'representative' of the physical state of the pair of particles, namely, the projection onto the singlet state. The complete state of the pair is therefore the singlet state.[14]

What does the Copenhagen variant say about a subsequent EPR–Bohm experiment, performed on the pair of particles? It seems that before the experiment, the Copenhagen variant does not attribute a state to each particle individually, but only to the 'measured system', which is the pair of particles. Therefore there is no sense in which we can speak of the evolution of the individual particles from the beginning to the end of the experiment. However, it then appears that we cannot determine whether the condition of independent evolution holds in the EPR–Bohm experiment.

How to proceed depends on a further issue. Consider the pair of particles at the end of the EPR–Bohm experiment, and suppose that the measurements on them have been non-disturbing. Suppose that the results of the measurment were $D^{\alpha}_{i,+1}$ and $D^{\beta}_{j,+1}$. Does the Copenhagen variant attribute to α the physical state $|i, +1\rangle^{\alpha}$, or does it only attribute to the composite system the state $|i, +1\rangle^{\alpha} \otimes |j, +1\rangle^{\beta}$?

If the former, then it is easy to show that the condition of independent evolution fails in the Copenhagen variant. Hardy has shown how to create an entangled state from two systems initially in a product state.[15] (I discuss Hardy's experiment, and the use he makes of it, in section 9.6.2.) Therefore, suppose that the Copenhagen variant has assigned to each of two systems a physical state, due to a previous measurement. Then use Hardy's procedure to create an entangled state and proceed (as Hardy describes) to do an experiment similar to the EPR–Bohm experiment, in which the outcomes are correlated, but space-like separated, and not screened off from each other by the initial state. Then the Copenhagen variant will assign to each particle the state corresponding to the outcome of that particle. The state assigned to one particle will, in general, not be independent of the state of the other, and therefore the 'evolution' of one particle from its initial state to its final state is not independent of the state of the other.

However, if only the composite system (the pair of particles) has a state at the end, then this scenario does not contain a violation of independent evolution. Moreover, it is difficult to imagine one that would. If it is the general policy of the Copenhagen variant to consider joint measurements of space-like separated local observables on entangled systems to be a single measurement of a property of the whole, then independent evolution will never fail, because there will never be the states L_α and L_β to discuss in the first place. (That is, separability fails.) Of course, it is hard to see why anybody would adopt this policy, except to avoid having to 'admit' a failure of independent evolution. Moreover, it is hard to see how the Copenhagen variant under this guise will be able to attribute states to the individual apparatuses — these too are entangled. In any case, there *is* the undeniable correlation between the two apparatuses, and no initial complete state as assigned by the Copenhagen variant will screen either apparatus off from the other.

Thus the (most reasonable form of the) Copenhagen variant appears to violate independent evolution. What about Lorentz-invariance?

The Copenhagen variant makes use of the concept of a 'measurement' to define the set of definite-valued events, and some will find this use unattractive. However, one advantage is that the concept, however untidy, carries over without harm to the relativistic domain. That is, whatever 'measurement' means in non-relativistic quantum mechanics, presumably it means the same in relativistic quantum mechanics. In that case, there is no problem in principle with a Copenhagen variant for relativistic quantum mechanics.

However, such an interpretation is not guaranteed to sit well with relativity. Consider first the version of the Copenhagen variant that endorses collapse. There, problems with relativity emerge as a result of the collapse of the statevector.

However, a version without collapse does no better. The hope that it would might arise from the belief that it is the collapse of the wave function that produces the difficulties with relativity. However, it is not quite collapse that produces the problem. Rather, the problem comes from the supposition that properties whose probability are less than 1, as given by a 'precollapsed' wave function, are nonetheless occurrent. Whether one then collapses the wave function is irrelevant. Put differently, the theoretical state transforms properly, in relativistic quantum mechanics, but that fact does not guarantee that the physical state does.

The fact that it does not, in the version of the Copenhagen variant without collapse, can be seen in the figures from section 9.2. Those figures were used to illustrate why CSL has trouble with Lorentz-invariance, but they serve here as well, for the CSL collapse is onto one of the possible physical states of the Copenhagen variant. Hence, as for CSL, different hyperplanes (or observers in different Lorentz-frames) will assign different physical states to the very same points of space-time.

However, it is not clear whether van Fraassen should be worried by any of these apparent problems. His attitude is avowedly empiricist, and there is no *empirical* trouble looming here. No experiment will bring the Copenhagen variant into empirical conflict with special relativity, because of the no-signalling theorem. If empirical adequacy is enough, then perhaps the Copenhagen variant is enough.

9.3.3 The Kochen–Dieks–Healey interpretation

Of course we know that some are not satisfied with empirical adequacy of van Fraassen's sort. I have already noted that authors such as Dieks and Healey want to be able to tell a story about what happens to quantum-mechanical systems when they are not being observed. They are also more concerned about non-locality than van Fraassen appears to be.

Both Dieks and Healey have considered in detail the implications of their interpretations for non-locality. A full account is not possible here. Hence, at the risk of injustice, I shall make only occasional mention of their detailed accounts, and focus instead on the consequences of chapters 6 and 7 for the Kochen–Dieks–Healey interpretation.

We have already seen that this interpretation is indeterministic, and we are therefore led immediately to suppose that it must be non-local in some sense. Nonetheless, it helps to go through the details. To that end, consider the standard EPR–Bohm experiment. In that case, the pair of particles begins in the singlet state. The reduced state for each particle is just the identity operator (ignoring the spatial degrees of freedom), and therefore the particles individually have only the trivial property, which is insufficient to determine the result of the measurement. Nor is the property possessed by the composite system as a whole enough, for that property is the projection onto the singlet state. Therefore, given the initial state of the particles, both individually and as a composite system, the various possible results of the measurements are not determined. The condition of deterministic results fails. The condition of deterministic transitions fails as well, as I discussed in chapter 4. Whether the condition of weak deterministic transitions fails will apparently depend on the definition of the equivalence classes prior to the time of the outcomes and on the details of the dynamics during the process of measurement. But we need not worry about these details, for the failure of the condition of deterministic results is enough to force the issue once again: one or both of the conditions of local determination and independent evolution must fail. As before, the former holds and the latter fails.

To see why, it is best to have in mind the complete state of the apparatuses-plus-particles at the beginning and end of the (ideal, non-disturbing) EPR–Bohm experiment. At the beginning (or, just prior to the interaction), the state is:

$$|\Psi_{EPR}(t_0)\rangle = \frac{1}{\sqrt{2}}\left(|z,+\rangle^\alpha|z,-\rangle^\beta - |z,-\rangle^\beta|z,+\rangle^\alpha\right)|\varphi_0\rangle|\chi_0\rangle, \qquad (9.1)$$

where $|\varphi_0\rangle$ and $|\chi_0\rangle$ are 'ready-states' of the apparatuses. After the interaction, the state is:

$$|\Psi_{EPR}(t_{final})\rangle = \frac{1}{\sqrt{2}}\left(|z,+\rangle^\alpha|z,-\rangle^\beta|\varphi_+\rangle|\chi_-\rangle - |z,-\rangle^\beta|z,+\rangle^\alpha|\varphi_-\rangle|\chi_+\rangle\right), \quad (9.2)$$

where $|\varphi_\pm\rangle$ represent the indicating states for one apparatus, and $|\chi_\pm\rangle$ for the other. (I have suppressed reference to the directions in which the spin is measured.)

One might think, because of (9.2), that the condition of local determination fails. Although the complete state of the space-time region where α is measured includes the state of the apparatus in that region, according to (9.2) the apparatus for α is *not* in an indicator state. Instead, its state is the projection onto the subspace spanned by $|\varphi_+\rangle$ and $|\varphi_-\rangle$. However, this result is an artefact of the unrealistic assumption that the measurement is ideal and that the environment plays no role in the interaction. As has been discussed at length elsewhere, when the idealizations are dropped, the reduced state of the apparatus is 'approximately' the 'correct' one.[16]

However, it is clear from (9.1) that the condition of independent evolution fails, even in realistic cases. In fact, because (9.1) makes no assumptions about the measurement, it is clearly a realistic description of the initial state.

From (9.1) we find that the physical state of each particle is the identity, while the physical states of the apparatuses are $|\varphi_0\rangle$ and $|\chi_0\rangle$. Hence for times, t, prior to the interaction, the state $L(t)$ contains this information. However, this information is clearly not enough to screen $D^{\alpha}_{i,\pm}(t^{\alpha})$ off from $D^{\beta}_{j,\pm}(t^{\beta})$. Even if we add to $L(t_0)$ the physical states of the various composite systems (for example, the system consisting of the pair of particles), the condition of independent evolution fails.[17]

Advocates of the Kochen–Dieks–Healey interpretation are prepared to admit non-locality, but they have apparently taken solace in various metaphysical doctrines. Dieks, for example, gives two points at which the non-locality of this interpretation becomes, for him, acceptable. I will consider those two points here, as well as some claims made by Healey in his extensive discussions.[18]

First, Dieks says that 'there are clearly no changes at a distance in already determinate physical quantities'.[19] He is right, because the condition of independent evolution does not fail if the spectral resolution of the reduced state of a system is not changing in time.[20] Only when the definite-valued events are changing in time can the condition of independent evolution fail, but how much comfort is this fact to those who are already worried about non-locality? Even worse, as Dieks admits, not only does the (newly determinate) property possessed by a system depend on the states of other systems, but also which properties are determinate in the first place depends on the states of other systems. If you were bothered by space-like separated events influencing a change from one determinate property to another already determinate property, then probably you are no less bothered by space-like separated events influencing a change from one determinate property to some newly determinate property, not to mention space-like separated events influencing a change in which properties are determinate in the first place.

Second, Dieks continues:

there are no nonlocal causal influences.... But there are correlations which are "holistic" in the sense that they are well-defined properties of the total system in spite of the fact that they are not built up from well-defined local properties.[21]

The idea seems to be that the correlations between the results of the measurements are not to be explained in terms of non-local action, but rather in terms of a holistic property of the combined system that somehow enforces the correlations between the subsystems. Presumably, the undeniable mathematical violation of locality is not evidence of 'non-locality' but of 'holism'.

There are plenty of reasons to be disturbed by holism — I gave some of mine in chapter 7. However, granting the doctrine, is it helpful? Or explanatory? I do not see that Dieks' version is. I can almost believe that there are events for a composite system that are not composed of events for the parts. Moreover, given such 'holistic' events, their occurrence would be well explained in terms of holistic properties or states. However, the outcomes of the EPR–Bohm experiment are manifestly not of this sort. They are local events, witnessed in the usual way as states of regions of space-time. It is hard to see how a holistic property can explain the occurrence of these local events, even on the terms of the modal interpretation itself, in which the local event is an element of the spectral resolution of the *reduced* state of a single system.

One might reply that although the holistic properties do not explain these local events, they do explain some other, holistic, event: the correlation. The local events, then, are explained purely in terms of local properties and states, while the correlation is explained in terms of the holistic property. However, apart from doubts that a correlation is an event, this response seems to lead to circularity. What is the holistic property that explains the correlation? What answer is there besides 'a correlation'? Moreover, the correlation itself seems to be completely reducible to localized events in the two regions. Or in any case, any change in the correlation requires a change in the localized events (the outcomes of the measurements), but then we are back to saying that the *holistic* property ('the correlation') can influence the occurrence of *local* events, *viz.*, the result of a measurement on a single, localized, system. Therefore, I cannot see that Dieks' holism buys him anything.

Healey is a bit more careful. He admits that *part* of the explanation of the correlation must be some sort of non-local 'connection' between the local events, though he also thinks that the holistic property of the composite plays an essential (non-circular) role in the explanation. Then, having admitted this non-local connection, he argues that it is nonetheless compatible with special relativity.[22]

I agree. I think that non-locality is *compatible* with special relativity. However, how comfortably does the Kochen–Dieks–Healey interpretation sit with special relativity as it is standardly interpreted? As for CSL, special relativity seems to make the picture offered by the Kochen–Dieks–Healey interpretation an odd one indeed.

Dieks fully admits the truth, namely, that the properties one ascribes to a system will in general depend on the hypersurface from which one ascribes those properties.[23] Dieks therefore seems to have adopted the 'hyperplane dependence' advocated by Fleming,[24] in which physical reality itself is relativized to a hyperplane. A point of space-time does not simply have a property, but has one property *as* a point on hyperplane and another (possibly incompatible) property *as* a point on a different hyperplane.

Why is the Kochen–Dieks–Healey interpretation forced to such lengths? Here is an intuitive account. Consider two hypersurfaces (such as σ_0 and σ_0^* in figure 9.1) having some region, ρ, in common, and consider the statevectors (as given by relativistic quantum mechanics) on each of σ_0 and σ_0^*. They will differ, in general. Now consider a system located in ρ. The reduced state of this system will in general depend on which statevector one uses. However, that dependence means that the possible properties for the system, and *a fortiori* its actual property, depend on which wave function one uses to determine those properties.

Healey seems to think that his modal interpretation is not forced to this position. I disagree. His view seems to be based on a consideration of two different hyperplanes in the EPR–Bohm experiment, one to the past of both outcomes, the other to the past of the outcome for β, but to the future of the outcome for α. In both cases, the reduced state of β is just the identity, and Healey concludes that '[β] has exactly the same dynamical [physical] properties after M_A [the measurement on α] as it did before M_A'.[25] However, Healey has fixed on a special case where the reduced state of one system is independent of the state of the other. *In general* the reduced state of a system depends on the entire wave function, and therefore in general the reduced state in a region of space-time will depend on from which hyperplane one calculates the reduced state. Again we see the point that while the theoretical state may be covariant, physical states derived from it need not be.

Indeed, careful analysis (which, among other things, does away with the assumption that systems must be localized inside some region of space-time) bears this intuitive argument out. It has been shown[26] in detail that the Kochen–Dieks–Healey interpretation is committed either to hyperplane dependence or to the existence of a preferred frame. (The latter avoids contradiction of the sort mentioned above because in this case, only *one* hyperplane through any given event may be used to calculate the state of a system at that event.) Therefore, either observers must relativize the physical state to a hyperplane, or there must be a preferred class of hyperplanes. Either way, although the move to a relativistic Kochen–Dieks–Healey interpretation appears to be possible, that move involves controversial doctrines of its own.

9.3.4 Bub's interpretation

As I noted in chapter 4, Bub's interpretation can go either way with determinism. This fact makes the discussion of locality a difficult issue. I have no definite conclusions, but only conditional ones, and they are already familiar from the discussions above. If Bub chooses an indeterministic dynamics, then there must be some form of non-locality in his interpretation. If not, then maybe there need not be any non-locality. The possibility of a local version of Bub's interpretation is best left for the next section, where I discuss this possibility in detail for Bohm's theory (which, recall, is a deterministic instance of Bub's interpretation).

The issue of Lorentz-invariance is extremely complex in Bub's interpretation, largely because of the difficulty of determining (in practice, not in principle) the determinate sublattices for arbitrary subsystems of the universe. My view is that any reasonable version of Bub's interpretation (i.e., one that resolves the measurement problem in a plausible way) will fail to be Lorentz-invariant. Preliminary analysis seems to support this view.[27]

9.4 Determinism and locality in Bohm's theory

9.4.1 Is Bohm's theory local?

9.4.1.1 The Laplacian ideal and locality in our universe

Technicalities aside,[28] Newtonian mechanics is a fully deterministic theory. The notion of probability does not enter the theory at any fundamental level. This fact about Newtonian mechanics led to Laplace's well-known vision of a world whose evolution could be predicted with complete accuracy by a sufficiently powerful demon using only the initial conditions and Newton's equations.

Quantum mechanics destroyed this vision, or so we learn. However, like it or not, Bohm's theory restores it. Bohm's theory is deterministic in all of the senses defined in chapters 6 and 7. Just as in classical mechanics, the notion of probability does not enter the theory at a fundamental level.

However, we must be careful about the sense in which Bohm's theory obeys the various forms of determinism defined in chapters 6 and 7. Although it is obvious from one point of view that Bohm's theory is completely deterministic — the evolution of the configuration of the universe is given by a deterministic equation — it is not immediately obvious how Bohm's theory satisfies the various forms of determinism defined in chapters 6 and 7.

To begin, what are the complete states, $L_n(t)$? I have been calling them 'states of the particle' but recall that they are states of time-slices of the backwards light cones of the outcomes. Therefore they will include information not only about the position of the particle, but also information about all of the particles in the measuring apparatus (or at least those involved in the outcome). Now consider a universe consisting of just the measuring apparatuses and the particles. In this case, $L(t_0)$ may be taken as the initial state of the universe, when the pair of particles and the apparautses were created. Given the Hamiltonian for this universe, one could, in principle, predict everything from the initial state, $L(t_0)$, including when the measurements will be made, and what measurements they will be. (Therefore, as I mentioned in chapter 7, the time of measurement is determined by the complete initial state, at least in this imagined universe.)

In particular, the conditions of deterministic transitions and deterministic results both hold in this universe. The former holds because all of the particles whose state are given in $L(t_0)$ evolve deterministically, and no additional particles will be described by $L(t)$ for some $t > t_0$, primarily because there *are* no additional particles in this universe, and anyhow if there were, still no additional particles can enter the light cone.[29]

The condition of deterministic results holds because the initial state $L(t_0)$ fixes the results of the measurements.[30] However, note that because the time and direction of the measurements are fixed by $L(t_0)$, there are two 'kinds' of 0 on the right-hand side of equation (7.5). One kind of 0 appears because the direction, d, is not the direction in which the measurement will occur, or because the time, t^n, is not the time at which the measurement will occur. In such cases, it is impossible for the detector to 'indicate' spin-r in the d-direction at time t^n, and therefore $p^{\lambda_0}(d_n^t) = 0$ for any d_n^t. The other kind of 0 appears because although the detector *will* indicate spin in the d-direction at time t^n, it will not indicate spin-r.

Now consider the locality conditions. Apparently all of the dynamical locality conditions fail. To see why, recall from chapter 5 that the expression (5.9) for the velocity of one particle depends on a wave function. In general, this wave function will refer not only to the particle in question, but also to others. More precisely, whenever the particle is in an entangled state, its velocity generally depends on the positions of the particles with which it is entangled.

In the case of a measurement of spin, Bell has derived an explicit expression for the trajectories of two particles in an entangled state.[31] It is clear from Bell's expression (given in note 32) that even when the particles are space-like separated, the instantaneous velocity of one can depend on the position of the other.[32]

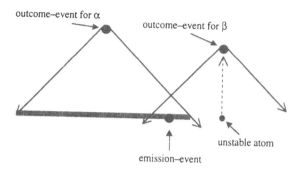

outcome–event for α

outcome–event for β

unstable atom

emission–event

Fig. 9.4. *Spin measurement in a simplified universe.* The unstable atom will either decay or not by t_β, and although it is determined in Bohm's theory whether it will, the information given by $L_\alpha(t_0)$ (depicted by the thick shaded line) is insufficient to make the determination.

It seems to follow that the condition of independent evolution fails. The evolution of $L_\alpha(t)$ *does* depend on the state of β, and therefore on $L_\beta(t)$. Moreover, conditionalization of the expectation value of $L_\alpha(t)$ on $L_\alpha(t_0)$ does not screen β off from $L_\alpha(t)$, because $L_\alpha(t_0)$ does not contain enough information to determine the evolution of β.

To see why this last statement is true, consider the following example. Suppose that the direction in which the apparatus for β will measure spin is determined by whether a certain atom has decayed. (See figure 9.4.) Suppose further that this atom is in the backwards light cone of the outcome for β, but not for α. As we saw in chapter 5, the velocity field determining the motion of β is partly determined by the direction in which the measuring apparatus for β measures spin. Hence, the motion of β is partly determined by whether the atom has decayed or not. However, the motion of α is partly determined by the position of β, and therefore by whether the atom has decayed or not. Therefore, because the state of the atom is not included in $L_\alpha(t)$ for any $t \geq t_0$, the condition of independent evolution must fail. That is, even though in Bohm's theory the time of decay for the atom is *determined*, that information is not contained in $L_\alpha(t)$ for any $t \geq t_0$, and therefore $L_\alpha(t)$ is not sufficient to fix the evolution of α, which depends on whether the atom decays. (In the figure, I have shown the region described by the state $L_\alpha(t_0)$ in some reference frame. In other reference frames different regions will be described, but this fact changes nothing about the argument.)

On the other hand, the condition of local determination holds. Although the motion of α depends on the state of β, once the position of α is known at time t_α (the time of measurement for α), the result of the measurement on α is fixed, and the position of β is irrelevant. Therefore, whether an event

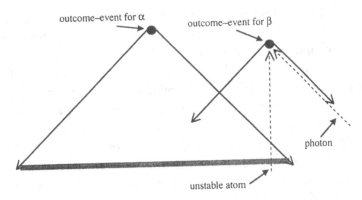

Fig. 9.5. *Spin measurement in a slightly more realistic universe.* Again, the unstable atom will either decay or not by t_β, but now the 'initial time' of the universe is sufficiently prior to the measurements that the backwards light cone of the outcome for α contains enough information to determine whether the atom will decay, and thereby to determine its eventual effect on the evolution of α. On the other hand, the photon is *never* in the backwards light cone of the outcome for α, and therefore $L_\alpha(t)$ never contains information about the photon.

occurs in a given continuously connected region of space-time is determined entirely by the state of that region, regardless of entanglements.

However, does Bohm's theory *really* entail the failure of the condition of independent evolution? There is no denying that *in this model*, i.e., in this scenario in which the universe is created at time t_0 and is, at that time, made up of just two particles in a singlet state and two measuring apparatuses, independent evolution fails in Bohm's theory, but of course, such a universe is not ours. Let us begin to make the scenario more realistic by supposing that, rather than having been created at t_0, the apparatuses and particles, including the unstable atom, have been around for a long time. In that case, it remains true that $L_\alpha(t_0)$ (in the reference frame as depicted in figure 9.4) is insufficient to determine whether the atom will decay or not, but now this fact is an accident of our definition of t_0. For example, we may take as the 'initial time' of the experiment t_{init} (see figure 9.5 — again the region described by $L_\alpha(t_{init})$ is shown only for a given reference frame.). It could be that, in this case, $L_\alpha(t_{init})$ *is* sufficient to determine whether the atom decays, and therefore sufficient to determine the motion of β, and therefore sufficient to determine the motion of α. If so, independent evolution will hold, replacing t_0 (in the statement of independent evolution) with t_{init} (and there is certainly no reason not to make this replacement).

However, let us move to the obvious next step. Consider in figure 9.5 a photon that travels at the speed of light. Suppose now that its polarization determines what measurement will be made on β. Suppose further that there was no beginning of the universe, so that this photon has been travelling at the speed of light from time $t = -\infty$. In that case, no matter how far back in time one goes, the backwards light cone of the outcome for α will not contain enough information to determine the polarization of the photon. Hence it appears that, for any t_{init} we choose and for any reference frame we choose (because time-slices cannot be time-like), $L_\alpha(t_{init})$ will not contain sufficient information to determine what the measurement on β will be.[33] Is this case a decisive proof of non-locality in Bohm's theory?

On the one hand, there is no denying that Bohm's theory has models in which this scenario occurs. Is our universe one of them? The prevailing opinion in favor of big bang theories seems to say 'no'. In the big bang theories, the initial state of the universe is in the backwards light cone of every point of space-time. *A fortiori*, $L_\alpha(t_{init})$ (where t_{init} is the time of the big bang) contains enough information to determine what the trajectory of β will be. Hence in our universe, it looks as though we are in the following situation in Bohm's theory: the conditions of local determination and independent evolution hold (when we substitute t_{init} for t_0 in the statement of independent evolution). In the next subsection, I discuss the relation between these points and Bell's theorem. In the subsequent subsection, I discuss whether, despite these points, Bohm's theory should nonetheless be considered non-local in some important sense.

However, first note that we would not have got this far if Bohm's theory had not been deterministic. Return to figure 9.5, and imagine now that the condition of deterministic transitions fails in Bohm's theory.[34] In that case, even though $L_\alpha(t_{init})$ includes complete information about the state of the photon at t_{init}, this information is not necessarily sufficient to determine the polarization of the photon at later times. Therefore, the argument for independent evolution above would fail.

9.4.1.2 Bohm's theory and Bell's theorem

Suppose Bohm's theory does turn out to be local. What then do we make of Bell's theorem? Typically locality is taken to be inconsistent with quantum mechanics because of Bell's theorem. I have already said why this claim is false, and Bohm's theory provides an explicit example of how the failure of λ-independence can help a local theory to escape the grasp of Bell's theorem.

First, why does Bell's theorem need λ-independence? Let us recall briefly Bell's derivation. Because Bohm's theory is deterministic, we may restrict attention to Bell's original derivation for deterministic theories. In that case, Bell assumes the existence of two functions, $A(i, \lambda)$ and $B(j, \lambda)$, which determine the result of the measurements on α and β respectively, given that spin was measured in the i-direction and j-direction respectively, and that the complete state 'at some suitable instant' was λ. Clearly the very form of these functions includes the assumption of factorizability, as Bell noted.

For which 'suitable instant' do the λ describe the state? It is most instructive to assume that they describe the state at the initial time, t_{init}. (The arguments for this case carry over with minor modification to the case where the λ describe the complete state just before the time of measurement.) Hence in Bohm's theory, the function $A(i, \lambda)$ may be written

$$A(i, L(t_{init})) = \sum_{r_\alpha = \pm 1} r_\alpha p^{L(t_{init})}(i_\alpha). \tag{9.3}$$

Using these deterministic functions, Bell writes down the following expression for the average value of the product of the results (a measure of their correlation):[35]

$$C(i, j) = \int_\Lambda A(i, \lambda) B(j, \lambda) \rho(\lambda) d\lambda. \tag{9.4}$$

Using the strict correlations, Bell notes that $A(a, \lambda) = -B(a, \lambda)$ for any a. Therefore, he is able to write

$$C(a, b) - C(a, c) = \int_\Lambda \left[A(a, \lambda) A(c, \lambda) - A(a, \lambda) A(b, \lambda) \right] \rho(\lambda) d\lambda. \tag{9.5}$$

From here one easily gets

$$C(a, b) - C(a, c) = \int_\Lambda A(a, \lambda) A(b, \lambda) \left[A(b, \lambda) A(c, \lambda) - 1 \right] \rho(\lambda) d\lambda \tag{9.6}$$

and it is a short step to Bell's inequality:

$$|C(a, b) - C(a, c)| \leq 1 + C(b, c). \tag{9.7}$$

However, in writing down (9.5), and more evidently in the move to (9.6), one implicitly assumes that λ is independent of the apparatus-settings. If it is not, then the correct form for (9.5) is

$$C(a, b) - C(a, c) = \int_{\Lambda_{a,c}} A(a, \lambda_{a,c}) A(c, \lambda_{a,c}) \rho(\lambda_{a,c}) d\lambda_{a,c}$$
$$- \int_{\Lambda_{a,b}} A(a, \lambda_{a,b}) A(b, \lambda_{a,b}) \rho(\lambda_{a,b}) d\lambda_{a,b}. \tag{9.8}$$

That is, we must *index* the λ with apparatus-settings, and then it is not possible to get from (9.8) to an equation similar to (9.6). Therefore, without λ-independence, Bell's inequality cannot be derived.

Now recall that in Bohm's theory, $L(t_{init})$ determines what measurements will be made on each particle. Therefore, $L(t_{init})$ is certainly not independent of the apparatus-settings. Therefore, Bell's inequality cannot be derived in this case. Moreover, it is clear that there is not (necessarily) any non-locality involved in this picture. The connection between the state $L(t_{init})$ and the apparatus-setting is not a non-local connection. Rather, it is just the connection that usually obtains between an initial and final state in a deterministic theory. Moreover, $L(t_{init})$ is (by definition) in the backwards light cones of the measurement-events.

Therefore, although Bohm's theory obeys independent evolution, and local determination, no Bell inequality can be derived, due to the failure of λ-independence. Bohm's theory may therefore be an example of the type of theory whose possibility I mentioned at the end of chapters 6 and 7: local, and consistent with quantum mechanics.

9.4.1.3 Contentious details: Determinism, causation, and counterfactuals

One could stop here, and emphasize what is perhaps a novel conclusion: Bohm's theory obeys the conditions of independent evolution and local determination, and it does so without being committed to Bell's inequality. Nor need the violation of Bell's inequality involve non-locality — it may only involve the failure of λ-independence due to the universal determinism of Bohm's theory.

In spite of these conclusions, surely most readers are still thinking, 'but it *must* be non-local in some sense'. Perhaps, but the point of this section is to show that whether Bohm's theory is non-local in any interesting sense is at the least a contentious issue. Behind this point is a more general one: I have included this lengthy discussion of the status of locality in Bohm's theory to illustrate one of the general contentions of this book, namely, that questions about locality, Lorentz-invariance, and the like cannot be decided by recourse to extremely general analyses, but must instead be discussed in the context of a detailed interpretation of quantum mechanics (or at any rate, in the context of a detailed — and dynamical — model of the EPR–Bohm experiment).

What reasons do we have left for thinking that Bohm's theory is non-local? The most obvious is the functional dependence of the velocity of some particles on the positions of space-like separated particles. Another is the supposed existence of a preferred frame in Bohm's theory, and the consequent lack of Lorentz-invariance at the level of the individual system. I shall return to Lorentz-invariance in the next subsection. Here I address the first worry.

A likely reaction to the argument of the previous two subsections is that despite the conclusions there, Bohm's theory is non-local, in the following sense: although there is in $L_\alpha(t_{init})$ enough information to specify what *will* happen in the future, and in particular to specify what measurement will be made on β and what the result will be, these events do not actually happen inside the backwards light cone of the outcome for α, and yet they have an *effect* on that outcome. In other words, the complete information in $L_\alpha(t_{init})$ is merely used to calculate the future non-local effect that the measurement on β will have on the motion of α.

Of course, this reaction begs the question against those who would argue for locality in Bohm's theory. The issue under contention is whether there *is* any 'non-local *effect*' of one particle on the other. The evidence for any such effect is the functional relation between the two. The issue therefore comes down to the status of these fuctional relations. Here again, one's initial reaction is likely to be that these functional relations indicate some form of non-locality: the velocity of one particle depends on the position of its space-like separated partner.

There is one obvious, but probably not convincing, reply. Normally a correlation of what happens in a region, ρ_α, with the state of a space-like separated region, ρ_β, is *not* taken to indicate non-local action if some event in the backwards light cones of ρ_α screens the state of ρ_β off from what happens in ρ_α. If you send two identical letters, one to the Prime Minister of England, and the other to the King of Thailand, and they are opened at space-like separation, the strict and non-local correlation between the contents (more precisely, the contents' being seen by the recipients) is no surprise. Bohm's theory may present the same situation. The correlations between space-like separated events were 'arranged' from the start. In any case, they are certainly, in the technical sense, screened off from each other by $L(t_{init})$.

Why is this argument not conclusive? The reason is familiar from attempts to describe causality in terms of probabilties: two correlated events may be screened off from each other by their common past, but it does not follow that there is no causation between them. Consider the following scenario. You meet with two people, A and B, and give them the following instruction: if A says 'up', then B says 'down'. Further, you instruct A to utter the sequence 'up-up-down'. Then let them converse. The result of the conversation is completely determined ahead of time, and in particular the strict correlation between A and B is already present in the initial state (your instructions to them). However, it is at least plausible that the arrow of causality is from A to B. That is, B said 'down-down-up' *because* A said 'up-up-down'. (One might rather say that the initial conditions are also a contributing cause for this result, but the only question here is whether causation of any sort, partial or complete, exists between A and B directly.) Might the functional relations be the same way? Might it not be that we can use $L(t_{\text{init}})$ to predict all outcomes because we can use it to predict how each of the particles will *cause* the other to move?

That view is natural enough, but my purpose here is to show that it is certainly not inevitable. Let us consider again, from the philosopher's point of view, why we want to say that there is causation between A and B in the example above. The reason, I think, is that the scenario appears to support counterfactuals. That is, even though A was *instructed* to say 'up-up-down', if A had disregarded those instructions and had said instead 'down-up-down', then B would have said 'up-down-up'.

Popular analyses of causality are, of course, exactly along these lines. Lewis gives roughly the following account:[36] C causes E if and only if: if it had been that not-C, then it would have been that not-E. There are plenty of riders attached to this basic statement, but these need not concern us here.

There is no chance here of establishing beyond doubt whether the functional relations in Bohm's theory involve causation. What I shall do is to proceed along broadly Lewisian lines and show that whether the functional relations involve space-like causality is very contentious. Indeed, I shall suggest that on a Lewisian analysis, they do not. (One may, of course, take this result as evidence against Lewis!)

Do the functional relations between particles support counterfactuals? Let us consider a two-particle universe, where the particles are in an entangled state at some time, t. Their coordinates are q_α and q_β. At that time we have the following functional relations (see (5.9)):

$$
\left.
\begin{aligned}
v_\alpha^\psi(t) &= \frac{1}{m_\alpha} \text{Im} \left(\frac{\nabla_\alpha \psi(q_\alpha, q_\beta\,; t)}{\psi(q_\alpha, q_\beta\,; t)} \right), \\[2ex]
v_\beta^\psi(t) &= \frac{1}{m_\beta} \text{Im} \left(\frac{\nabla_\beta \psi(q_\alpha, q_\beta\,; t)}{\psi(q_\alpha, q_\beta\,; t)} \right).
\end{aligned}
\right\}
\tag{9.9}
$$

It is clear that, for example, in the expression for $v_\alpha^\psi(t)$ we could put in any number of values for q_β, thus, apparently, constructing counterfactuals of the form: if β had been at q_β, then the velocity of α at time t would have been $v_\alpha^\psi(t)$.

However, in principle, in Bohm's theory, to proceed along these lines is invalid. Recall from chapter 5 that probabilities in Bohm's theory enter only at the level of a distribution over possible initial configurations of the universe. When the position of a particle at time t is discovered, the proper thing to do in Bohm's theory is to go back to the initial conditions and select from all possible initial conditions (still given the same initial wave function) those that are compatible with the known location of the particle at t, and calculate everything over again. In *practice* it is not necessary to follow this procedure. Instead, in practice one can simply plug the discovered value for q_β into the equations (9.9) to find the velocity for α, and get the same answer. The point here, however, is a conceptual one — the latter procedure (i.e., substituting the discovered value for q_β into the equations (9.9)) works *only* because it is gives the same answer as the former one (i.e., going back to the initial conditions), which is the conceptually correct procedure to follow in a determinstic theory such as Bohm's theory.

To put it differently, and now in terms of counterfactuals, to assume that q_β has some value other than its true value (i.e., to make a counterfactual assumption about the location of β) is to assume that β has a location that is *incompatible* with the initial state of the universe — within Bohm's theory, the initial state entails one and only one location for β at t. There are two ways to deal with this problem. The first is to ignore it, i.e., to maintain that the initial conditions are the true ones, but to suppose counterfactually that the position of β is other than it is. Call this method of evaluation the 'non-backtracking method'. Using the non-backtracking method, the antecedent of the counterfactual is necessarily false within Bohm's theory — it requires us to imagine, in Lewis' words, a 'minor miracle',[37] a temporary suspension of the laws of Bohmian mechanics. Then even if the counterfactual turns out to be true, what have we learned about Bohm's theory? Can one show that there is causality of some sort in a theory by evaluating a counterfactual whose antecedent entails the falsity of that theory? I return to this question below, and argue that even if one *can* learn something about Bohm's theory from the non-backtracking method, it is not clear that it leads to space-like causation.

The second way to deal with the problem, the backtracking method, is to admit into the antecedent of the counterfactual not only the position q_β, but also the initial conditions that are entailed by it.[38] In this case, one does as I said above: namely, one selects from all possible initial conditions those that are compatible with the value of q_β and evolves this ensemble in Bohmian fashion to the time t to discover the velocity of α situated at q_α. Here, in general, one will find that a change in the position of β *does* result in a change in the velocity of α.

However, again we are faced with the question of whether this fact indicates space-like causation. The answer seems to be 'no', for apparently what is responsible for the change in the velocity of α in this case is the change in initial conditions.

To illustrate this point further, consider the following example. Suppose I cut an orange in half — label the halves 'O_1' and 'O_2' — and I give O_1 to you. Now I ask: Which half *would* you have had if I had had O_1? Let us evaluate this counterfactual using both methods.

According to the backtracking method, the answer is 'You would have had O_2 (rather than O_1)'. But is there any causation between us (or, our halves) in this case? Strong advocates of the counterfactual analysis will say 'yes', but even they should admit that the causation is indirect. That is, although the truth of the counterfactual 'If I had had O_1, you would have had O_2' indicates some causal relation or other between O_1 and O_2, the relation in this case is completely accounted for by a common cause. In Bohm's theory, too, the backtracking method yields a true counterfactual: 'If β had been elsewhere, then the velocity of α would have been otherwise'. However, the *reason* that the velocity of α would have been otherwise is that the new position of β entails a different initial condition, which in turn entails a different velocity for α.

What about the non-backtracking method? I have already indicated that it seems to me less capable of teaching us anything about Bohm's theory, but let us consider it again nonetheless. In the example, the non-backtracking method can be applied in two ways. To see what they are, it is easiest to use Lewis' method for evaluation of counterfactuals. Denoting the counterfactual 'If it were that φ it would have been that χ' by '$\varphi \,\Box\!\!\rightarrow\, \chi$', Lewis proposes:

'$\varphi \,\Box\!\!\rightarrow\, \chi$' is true at a world w iff: (i) there are no φ-worlds (i.e., possible worlds where φ is true) or (ii) *every* φ-world among those worlds closest to w is also a χ-world.[39]

Condition (ii) is the important one here. Which of the worlds in which I keep O_1 are closest to this world (where I give O_1 to you)? Probably the intuitive answer is the worlds where you get O_2, but remember that we have postulated a minor miracle, and have not changed the initial conditions. In that case, it is less clear. Is the nearest world the one in which the strict anti-correlation between your half and mine is violated (and thus you have O_1 also, as the initial conditions imply), or is the nearest world the one in which the correlation is preserved, and a second minor miracle occurs, in which case you have O_2 despite the initial conditions?

The answer to this question is not at all obvious. Nor are the consequences for Bohm's theory obvious. If you choose to say that the functional relations may be violated in order not to postulate a second minor miracle (the change in the velocity of α), then the counterfactual analysis yields no causation.

If you choose to say that the functional relations must not be violated in moving to the counterfactually nearest world, then you will find that a second minor miracle occurs, and along with the miraculous change in the position of β, there is a miraculous change in the velocity of α. The strict adherent to the counterfactual analysis will say that there is some form of causation here, but the preponderance of miracles makes it difficult to know what to conclude. Does the occurrence of two space-like separated but correlated miracles indicate the presence of space-like causation?

There is plenty more to ask about Bohm's theory. For example, how does Bohm's theory fare according to other analyses of causality? It is a central feature of Salmon's analysis,[40] for example, that causal processes propagate continuously in space and time. In that case, it is obvious that the functional relations will not be an indicator of causes, though other difficult questions would no doubt arise, some of them about locality. For example, if Bohm's theory is local — if all (Salmon-)causes of an outcome are in its backwards light cone, and the functional relations are short ways to tell the longer causal story — then is it non-Markovian? The reliance on initial conditions in principle to calculate everything suggests that it might be — the initial state of a light cone might determine the state inside the light cone at later times, even though the information contained in the initial state is not contained in the complete state of a time-slice of the light cone at intermediate times. That is, if we say that events outside the backwards light cone of an outcome do not influence the outcome, then we might be forced to say that the outcome somehow 'remembers' the entire initial state, even though not all of the information in the initial state propagates continuously to the outcome. These questions and many more are still open in Bohm's theory.

Will the result of examining these questions in detail be to conclude that Bohm's theory might be local? Perhaps, but I think that probably the proper lesson is that the very notion of locality — perhaps even the notion of causality — does not have application in a completely deterministic world. The problems that arose for the application of these concepts did so because Bohm's theory is fully deterministic. In an indeterministic theory, there will not, in general, be some initial state of the universe that fixes whether any given event will occur. Nor will the problems with the evaluation of counterfactuals arise. Note also that although I did sometimes rely on the special initial conditions of our universe (the big bang), many of the points in this section are independent of initial conditions. If a Bohmian universe does not permit an interesting notion of causation, for example, then presumably it does so regardless of initial conditions.

Now one might object to this suggestion that it is exactly the cases of *deterministic* causation that we understand best, while so-called 'probabilistic causation' is the hard case.[41] I agree, but there is a difference between 'determinsitic causation' and 'causation in a determinsitic world'. The former is causation where the cause invariably acts so as to produce its effect, and it can happen even in an indeterminsitic world. I am suggesting that even deterministic causation might not make sense in a completely deterministic world.

Much more work needs doing. It may be that even in a universe in which every event is fated to happen from the start, some interesting notion of causality or action-at-a-distance can be formulated, but what does seem clear is that these will not be just our familiar notions. In the end, this conclusion is perhaps not so surprising. After all, the Laplacian ideal is metaphysically foreign to most of us; that apparently familiar concepts such as locality and causality should not be suited to this ideal is only to be expected.

9.4.2 Bohm's theory and relativity

9.4.2.1 The apparent problem: Hardy's argument

Another feature of Bohm's theory that 'everybody knows about' is its lack of Lorentz-invariance. Although non-locality gets more press in the philosophical literature, Lorentz-invariance seems to raise a more important issue. In any case, it raises a more or less clearly defined issue. As we saw in the last subsection, the question of whether Bohm's theory (or, indeed, any theory) is ('metaphysically') non-local easily becomes transformed into questions to which there may never be generally agreed answers. Lorentz-invariance, though slippery, is not as slippery as non-locality. It is fairly clear what it means for there to be no preferred frame, or for the hidden variables, or the empirical predictions, of a theory to have the correct transformation properties. (Of course, which of these conditions, if any, is *required* by 'Lorentz-invariance' is itself a difficult question.)

The usual claim made about Bohm's theory on this point is that it does have a preferred frame, and that the hidden variables lack the correct transformation properties. It is said further that in the statistical average Lorentz-invariance is recovered, so that only if we could measure the exact trajectories of particles (which, as we saw in chapter 5, is impossible) could we detect the preferred frame. However, conclusive arguments for a lack of Lorentz-invariance are not easy to find. One well-known attempt is the argument by Hardy.[42] In this subsection I summarize his argument, and in the next I say why it does not apply to Bohm's theory.

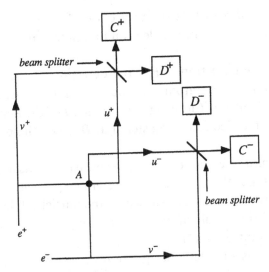

Fig. 9.6. *Hardy's gedankenexperiment.* The electron and positron enter the interferometers, and follow either the 'u' path or the 'v' path. If they both follow the u path then they meet at A and annihilate each other. Otherwise, they both reach a beam splitter, and from there go to either the C or the D detector.

Hardy imagines two interferometers, interlocked as shown schematically in figure 9.6. A positron, e^+, enters one interferometer, taking either path v^+ or path u^+ with equal probability. At the other end of the interferometer the paths meet and pass through a beam splitter, after which the positron registers either at detector C^+ or at detector D^+. The electron is in the same situation with respect to the v^- and u^- paths. If each particle takes its path u, then they meet at point A and annihilate one another. In this case there is no detection-event.

Hardy then considers three reference frames. In the first, F^+, the detection of the positron occurs before the electron has passed through its beam splitter. Hardy writes down a wave function in this case, $|\psi_{F+}\rangle$, which he derives simply by postulating the effect on the state of each particle of the various parts of the apparatus. He then assumes that the state of the positron has undergone the effects of the interferometer, the annihilation, and the beam splitter, whereas the state of the electron has undergone the effects of only the interferometer and the annihilation. Hardy argues that in F^+, if the positron is registered at D^+ then the electron must be on path u^-. This result is a little surprising, because one would expect that the registering at D^+ could occur as a result of any of three joint paths: v^+u^-, v^+v^-, or u^+v^- (but not u^+u^- because of the annihilation at A). However, Hardy has cleverly arranged things so that the projection of $|\psi_{F+}\rangle$ onto the positron's being at D^+ results in a state of the electron that puts it at u^- with probability 1.

In the second reference frame, F^-, the detection of the electron occurs before the positron passes through its beam splitter. Using an argument parallel to the previous one (or using the symmetry of the apparatus), Hardy says that in F^-, if the electron is registered at D^- then the positron must be on path u^+.

Finally, in the third reference frame, F, the measurements are simultaneous. In this case, the state of the system before either particle goes through its beam splitter is orthogonal to a state where the particles take the joint path u^+u^- (because of the annihilation). Therefore, in this case the particles cannot be on the joint path u^+u^-.

Now, Hardy points out that the question of whether a given particle is on a given path should have a Lorentz-invariant answer — if the positron is on u^+ for you, then it should also be there for me. Hence, consider an experiment in which the final result is that the detectors D^+ and D^- each registers a hit. In that case, an observer in F^+ will conclude that the electron must be on u^-. This statement must also be true for an observer in F^-, who concludes as well that the positron must be on u^+. Moreover, *both* of these statements must be true for an observer in F. However, they cannot be, because such an observer predicts with certainty that the particles cannot be jointly on u^+u^-. In other words, the assumption that the answers to questions of the form 'Is the particle on the path?' are Lorentz-invariant leads to a contradiction. However, in Bohm's theory, a particle *is* always on a given path, and therefore it must be that, in Bohm's theory, the answers to questions of that form are not Lorentz-invariant.

Hardy's contention, therefore, is that Bohm's theory cannot be made relativistic. Instead, he says, there must be a preferred frame. For example, the frame F^+ might be the 'preferred' frame, the frame of reference in which physical processes 'really' occur. In that case, it is not legitimate to compare the facts as they are in F^+ with what appears to be the case in other frames. What is the case is what is the case in F^+. Bohm's theory is (possibly) relativistic only at the statistical level.

Hardy's argument is potentially important because it does not rely on anything except the assumption that particles follow trajectories and some standard quantum-mechanical calculations. His conclusion echoes a common view that Bohm's theory cannot be made relativistic at the level of trajectories for individual particles.[43] The usual reason given is that because there are non-local connections, indeed instantaneous ones, there must be a preferred frame picked out by these instantaneous connections. That is, because the velocity of one particle 'at a time t' is determined in part by the position of others 'at the same time, t', there must, one would think, be a frame in which these times really are the same, a frame in which the positions of the other particles really do 'determine' the velocity of the particle. This frame is the preferred frame.

9.4.2.2 But is it really a problem? Reply to Hardy

It is exactly the generality of Hardy's argument that, in fact, renders it harmless to Bohm's theory. Once more we will see that a purportedly general analysis fails precisely because it attempts to be completely general.

In a comment on Hardy's argument, Berndl and Goldstein have shown where Hardy's argument falters.[44] As is explicit in Hardy's original formulation of the argument, to get any empirical contradiction from Hardy's scenario, to say that a particle is on the path $u^{\pm}$, must mean that it would be measured to be there, if the appropriate apparatus were present. However, recall from chapter 5 that Bohm's theory is contextual, in the sense that the path of a particle can depend on the setting (including the absence or presence) of apparatuses. Now look again at Hardy's argument. His claim that the occurrence of D^+ in frame F^+ entails that the electron is on u^- is true *in the context* of there being a detector on u^+ to measure whether the positron is there. Likewise for his claim that the occurrence of D^- in frame F^- entails that the positron is on u^+. Finally, his claim that the particles cannot be on the joint path u^+u^- is true in a context where detectors are present on both paths.

Results from one context cannot be compared with results from another context. One reason often given is non-locality — the presence of a detector on the u^- path non-locally affects the result of the measurement on the positron, and therefore there is no reason to expect that if the detector had not been on the u^- path, then the result D^+ would have been found. An equally good reason is determinism — the presence of the detector on the u^- path entails a certain set of initial conditions, and therefore there is no reason to expect that if the detector had not been on the u^- path (which means, if the initial conditions had been substantially different) then the result D^+ would have been found.

Where does that leave Bohm's theory as regards Lorentz-covariance? As I noted earlier (note 44), current Bohmian versions of relativistic field theories are non-covariant at the level of inidividual processes (though they are covariant at the level of statistical predictions, meaning that in practice the non-covariance cannot be detected). The weight of evidence, therefore, goes against the possibility of a completely covariant verson of Bohm's theory. However, it is worth noting that this evidence is not conclusive. To begin, we may note that even if one finds the arguments of the previous section unconvincing and believes that Bohm's theory is non-local, non-locality itself does not conflict with special relativity.[45] Nor does the existence of Bohmian trajectories by itself force non-covariance.

It is perhaps worth seeing this latter point in an example. It is obvious that the velocity field derived from the Schrödinger equation is not relativistic. To have a chance at a relativistic theory, we need to look at relativistic quantum mechanics, which means, most simply, the Klein–Gordon equation or the Dirac equation. There does not exist a satisfactory and general Bohmian interpretation of the Klein–Gordon equation;[46] hence we must look at the Dirac equation ($\hbar = c = 1$):[47]

$$\left(i\gamma^\mu \partial_\mu - m\right)\psi(x) = 0, \tag{9.10}$$

where $\psi(x)$ is now a four-component object (a bi-spinor) and the γ^μ are four-by-four matrices. Using the same techniques as in the non-relativistic case, one can derive from (9.10) a conserved 4-current:

$$j^\mu = \psi^\dagger \gamma^0 \gamma^\mu \psi. \tag{9.11}$$

This current is a time-like vector, and its time component is positive-definite. Indeed, $j^0 = \psi^\dagger \psi$ in every frame (because $\gamma^0 \gamma^0 = 1$). Therefore, one may define a Bohmian velocity field in the usual way:

$$v^i = \frac{j^i}{j^0} = \frac{\psi^\dagger \gamma^0 \gamma^i \psi}{\psi^\dagger \psi}, \tag{9.12}$$

where v^i is the ith component of velocity ($i = 1, 2, 3$).

As Holland has shown,[48] this 3-velocity follows from a straightforward definition of the 4-velocity, one that transforms as a normal 4-vector. That is, it is the velocity obtained by taking the 4-velocity to be tangent to the particle's world line.[49] Hence the Bohmian interpretation of the Dirac equation yields a purely relativistic expression for the velocity fields. If you use your wave function to obtain a velocity for a particle, and I use mine (a Lorentz-transform of yours), then our results will be Lorentz-transforms of each other.

Perhaps a simple example will help to illustrate. Consider a free positive-energy electron with helicity $+1$ moving with momentum p^3 (velocity p^3/E) in the laboratory frame. (NB, p^3 indicates momentum p in the x^3-direction.)

The wave function for this electron in its rest frame (the 'primed' frame) is:[50]

$$\psi'(\mathbf{x'}, t') = \frac{1}{\sqrt{V}} \sqrt{\frac{E+m}{2m}} \begin{pmatrix} 1 \\ 0 \\ 0 \\ 0 \end{pmatrix} e^{-imt'}, \tag{9.13}$$

where V is a volume of normalization. It is trivial to use (9.12) to find that the velocity of the particle in its rest frame in the x^3-direction is 0, as it should be.

Now consider the wave function as given in the laboratory frame:[51]

$$\psi(\mathbf{x}, t) = \sqrt{\frac{m}{EV}} \sqrt{\frac{E+m}{2m}} \begin{pmatrix} 1 \\ 0 \\ \dfrac{p^3}{E+m} \\ 0 \end{pmatrix} e^{ip^3 x^3 - iEt}. \tag{9.14}$$

It is not difficult to calculate the velocity of the electron in this frame using (9.14), because the constant terms cancel and the exponentials multiply to 1. The result (as Holland notes, in a different context[52]) is p^3/E. Of course, this result is exactly what we need: the velocity obtained from wave functions that were Lorentz-transforms of one another yielded velocities that are Lorentz-transforms of one another.

Admittedly, the case of the one-particle Dirac equaion teaches us very little. However, as soon as we turn to more complex cases, we run into two problems. First, there is not always a clear relativistic quantum theory in the first place (e.g., of a two-particle Dirac equation). Second, the failure to write down a covariant Bohmian version of an existing quantum theory is evidence, but not conclusive evidence, for the claim that Bohm's theory *must* be non-covariant.[53]

A general argument that Bohm's theory must be non-covariant might proceed as follows. Whether we take the functional connections between particles to be evidence of non-locality or a manifestation of determinism and initial conditions, they are present in Bohm's theory, and these connections make Lorentz-covariance impossible.

But do they? Correlations between events in space-time are not undeniable evidence of non-locality, much less of non-Lorentz-covariance. Instead, the correlations between space-like separated events may be seen as a consequence of the initial conditions. One need not assume that one particle 'determines' the velocity of another, and therefore one need not suppose that there must be some preferred frame in which this determination occurs.

Instead, there are just correlations, 'non-local', to be sure, but no different from (and no less relativistic than) familiar correlations due to a common cause.

The conclusion is that although Bohm's theory is *probably* non-covariant — in the sense that no empirically adequate Bohmiam-looking interpretation of relativistic quantum mechanics will be found that is covariant — there is (so far) no proof that such a thing cannot exist. This fact alone is important, for it tells against the usual story, which is that Bohm's theory, unlike other interpretations, is necessarily non-covariant.

9.4.2.3 Postscript: On the possibility of relativistic particle theories

Another argument has appeared, which one might take to show that in fact there can be no relativistic quantum theory of particles. The argument is due to Malament.[54] He assumes the existence of a position operator, places some reasonable restrictions on the 'quantum' theory in which this operator appears, and derives a contradiction.

The important restriction is that eigenprojections of the position operator corresponding to space-like separated regions commute with one another. This condition is meant to capture some version of Lorentz-invariance, and for our purposes, we may suppose that it does. To this condition, Malament adds a few other quite reasonable (though certainly not inviolable) conditions, such as that the energy be bounded below, and derives a contradiction. He concludes that there can be no relativistic 'quantum' theory of particles.

Of course, one can *define* a relativistic quantum theory of particles to be one that obeys Malament's condition, and then his conclusion follows trivially from the theorem (whose validity, and importance, is not in dispute here). Does it follow that the project of finding a relativistic version of Bohm's theory — or indeed any relativistic theory of particles that is empirically satisfactory at the quantum-mechanical level — is doomed?

No. Malament's conditions are far too strong to support that conclusion, for they presume that any 'quantum' theory must represent position with some operator. We have already seen that the Bohmian interpretation of the Dirac equation does not do so. Indeed, even the non-relativistic version of Bohm's theory does not *really* represent position with the position operator — it is merely a coincidence that the Bohmian position of a particle corresponds to its quantum-mechanical value for the position operator when position is measured. (Recall the discussion of 'naïve realism about operators' in chapter 4 — Bohm's theory does not presume that operators correspond to physical quantities but must instead *derive* this correspondence in the context of measurements.)

So apparently there are more ways to be 'quantum-mechanical' than to represent physical quantities with operators — or in any case, a theory can be empirically equivalent to standard quantum mechanics without adopting its mathematical formalism. However, then we must *not* take Malament's result as indicating the impossibility of a relativistic theory of particles that is empirically equivalent to standard quantum mechanics.

9.5 Probability, non-locality, and the sub-phenomenal world

I began this book (in the preface) by sidestepping the issue of realism, supposing (with all realists, and at least some anti-realists) that quantum mechanics should be read as *purporting* to tell us about the sub-phenomenal world — the world that is not accessible to our direct experience. That supposition leads immediately to the problems that have come up throughout this book, most significantly the measurement problem and the problem of how to reconcile quantum mechanics with the theory of relativity.

These problems are evidently related, and in several ways, two of which I hope became clear in this book. First, questions about locality should ultimately (I have argued) be considered in the context of specific proposal to solve the measurement problem, i.e., in the context of a detailed interpretation of quantum mechanics. General analysis can go only so far in helping us to understand how quantum mechanics and the theory of relativity are related. Second, and related to the first, understanding *how* a given interpretation solves the measurement problem — and in particular, whether and in what senses it is deterministic or not — is important for understanding the status of locality in that interpretation and its relation to the theory of relativity.

These points show the importance of the assumption that quantum mechanics purports to be about the sub-phenomenal world. Without that assumption, there is no need of interpretation — indeed, the very act of interpretation (as conceived here, at least) seems to presuppose that physical theory is about the sub-phenomenal world, for as I mentioned in the preface, standard quantum mechanics alone seems to be sufficient for one, such as Bohr, who is willing to let it apply only to the phenomenal world. So interpretation (or anyway, its importance) seems to presuppose that quantum mechanics purports to tells us about the sub-phenomenal world.

Therefore, if addressing the sorts of question addressed here requires interpretation, then it also requires the supposition that quantum mechanics purports to tells us about the sub-phenomenal world. It is here that we find, I think, the true importance of that assumption. Its importance has little to do with the debate about realism — which is itself (in my view) of little importance — but rather with its role in philosophy of physics: the questions that are of greatest interest in philosophy of physics presuppose that quantum mechanics is about the sub-phenomenal world (and, indeed, that there is no important distinction between the world of our everyday experience and the sub-phenomenal world). We do *not* here have a case of physics, or philosophy of physics, teaching us something about more general debates in philosophy of science. Instead, philosophy of physics has its own reasons for adopting certain general principles about science. At the same time, we *do* have here a case where those general principles cannot be ignored in the philosophy of physics.

However, as I have stressed all along, general principles cannot answer the questions that philosophy of physics asks. Indeed, they are required precisely *because* the questions that philosophy of physics asks must ultimately be addressed in the context of a detailed physical theory. Is quantum mechanics non-local? Is it compatible with the theory of relativity? *These* questions are meaningless, Bell's theorem, Hardy's argument, and the rest notwithstanding. Is Bohm's theory non-local? Is the Kochen–Dieks–Healey interpretation compatible with the theory of relativity? These questions are meaningful, but very difficult indeed. I have answered none of these meaningful questions here, but I hope to have pointed the way to what must be done to answer at least some of them.

Notes

Preface

1 See, for example, van Fraassen (1980).
2 Van Fraassen (1980, p. 72).
3 In addition proofs of most theorems in the book can be found in Dickson (1995d).

Chapter 1

1 There are several sources useful for gaining some familiarity with the formalism and approach to follow. See, for example, Beltrametti and Cassinelli (1981), Bub (1974), Gudder (1988), and Hughes (1989).

2 Actually, axiom (3) holds also when either of P or P' is contained in the other (as a subspace), or, in terms of projections, when either $PP' = P$ or $PP' = P'$. However, the point remains that axiom (3) is not completely general.

3 Gleason (1957) showed that every probability measure over a Hilbert space is given by a density operator, and vice versa, for any Hilbert space of dimension greater than 2. Density operators are generalizations of quantum-mechanical statevectors. For a good introduction to density operators, see Cohen-Tannoudji, Bernard, and Laloë (1977).

4 Lüder's rule assumes that $\text{Tr}[WP'] \neq 0$. If $\text{Tr}[WP'] = 0$ then it is still possible to define a conditional probability measure, but the definition is not unique, in the sense that there are many definitions that reduce to Lüder's rule when $\text{Tr}[WP'] \neq 0$, and there is apparently no reasonable criterion that would single out just one of these definitions. See Beltrametti and Cassinelli (1981, ch. 26), for some details.

5 See Bub (1977) and Hughes (1989) for discussion and proofs. The criterion is 'natural' because if P is contained in P', then (the event represented by) P' has probability 1 whenever P does.

6 In essence, one maps the sample space, Ω, from the classical space onto a set of mutually orthogonal projections that span some $\mathcal{H}$. Gudder (1979, pp. 20–21) gives a precise statement of the relevant theorem and proves it.

7 This claim follows immediately from a theorem due to Nelson (1967, Theorem 14.1). Some have tried to generalize the notion of a 'joint distribution' so that all observables do have a 'joint distribution', but these generalized notions, at the least, do not have some of the intuitive properties one would expect of a joint distribution. (See, e.g., Gudder (1979, pp. 88–94) and Beltrametti and Cassinelli (1981, p. 24).)

8 When the Hamiltonian is time-dependent, $U(t)$ is instead given by $U(t) = \exp[-i \int_0^t H(\tau)d\tau]$. The derivation of $U(t)$ from Schrödinger's equation is found in many textbooks. See, for example, Cohen-Tannoudji et al. (1977).

9 As far as I know, Schrödinger (1935a) was the first to consider this problem to be truly problematic. It is therefore sometimes referred to as the problem of 'Schrödinger's cat' (because Schrödinger's original example involved not a measuring apparatus, but a cat). Von Neumann (1955) (which was originally published in 1932) noted the problem, but took it to be resolved by his projection postulate — see chapter 2.

10 For a brief introduction to the tensor-product formalism, see Cohen-Tannoudji, Bernard, and Laloë (1977).

11 In the context of 'unsharp' quantum mechanics, this point must be made in a slightly more subtle way, for there it is possible to measure non-commuting but unsharp observables. See, for example, Busch and Lahti (1984).

12 The discussion to follow owes a great deal to the work of Wessels (1981) and Beller (1990). The reader is encouraged to consult these essays for a more thorough, and informed, account of the introduction of Born's rule. See also Heilbron (1988) and Jammer (1966, pp. 281–93).

13 Wessels (1981, p. 195).

14 Schrödinger's wave mechanics was introduced in his (1926). Beller (1990) discusses the influence that Schrödinger's theory had on Born. Born's preliminary report was his (1926a). His later paper on the same topic (discussed below) was his (1926b).

15 Wessels (1981) and Beller (1990) argue for these points. See, for example, Wessels (1981, p. 193).

16 Born (1926a).

17 Both quotations are from Born (1926b).

18 Pauli (1927).

19 Born (1927) later reiterated his addmission of the serious possibility of determinism.

20 Born (1926b).

21 Born (1926b).

22 Born (1926b).

23 For an extensive discussion of whether a guiding field of roughly the Bohmian type could have been (both sociologically and scientifically) accepted in the early development of quantum mechanics, see Cushing (1994b, esp. ch. 10).

24 However, Beller (1990, esp. section IV) argues that Born 'did not, in fact, have any definite, strong opinion on this issue [of indeterminism], but only tentative, uncommitted suggestions'.

25 Born (1926b).

26 Born (1926b).

27 Jordan (1927).

28 Jordan (1927).

29 The term was coined (as far as I know) by Fine (1970).

30 Those who are comfortable with talk about possible worlds may consider this account to be equivalent to Earman's (1986) preliminary account of determinism, which is roughly that a theory is deterministic if whenever two worlds agree about everything up to some time, t, they must agree at all times after t.

31 In terms of probabilities, the Markov property is that for all times $t_1 < t_2 < \cdots < t_n < t$,

$$p(S_t | S_{t_1}, S_{t_2}, \cdots, S_{t_n}) = p(S_t | S_{t_n}),$$

where S_{t_i} is the state at time t_i.

32 Given two states, σ and σ', time-homogeneity is the requirement that

$$p(S_{t+h} = \sigma | S_t = \sigma') = p(S_h = \sigma | S_0 = \sigma')$$

for every t and every h. ($S_t = \sigma$ means 'the state at time t is σ'.)

Chapter 2

1 The last equality follows from the spectral decomposition theorem and the linearity of $U(t)$. By the spectral decomposition theorem,

$$A(0) = \sum_i a_i P_{a_i}^{A(0)}$$

and therefore

$$
\begin{aligned}
A(t) &= U^{-1}(t)A(0)U(t) \\
&= U^{-1}(t) \left(\sum_i a_i P_{a_i}^{A(0)} \right) U(t) \\
&= \sum_i a_i U^{-1}(t) P_{a_i}^{A(0)} U(t).
\end{aligned}
$$

2 My discussion of the projection postulate is certainly not comprehensive. For more detailed discussions, see, for example, von Neumann (1955), Groenewold (1957), Margenau (1963), Teller (1984), and van Fraassen (1991).

3 See von Neumann (1955).

4 More generally, the condition for the results to match is that the operator representing the measured observable commute with the Hamiltonian that governs the evolution of the system between measurements. Even if it does not, there is a more general sense of 'repeated measurement', in which we measure the observable represented by A at time 0, then the observable represented by $U^{-1}(t)AU(t)$ at time t. In this case, if the result of the first measurement is P_a^A, then the result of the second measurement will be $U^{-1}(t)P_a^A U(t)$ with certainty.

5 There are plenty of other research programs along the same lines. For just some examples, see the following articles: Bell (1987a), Diosi (1992), Ghirardi, Rimini, and Weber (1986), Gisin (1984, 1989), and Primas (1990). Many additional references are listed in Dickson (1994b).

6 One often sees in discussions of CSL the field of processes $w(\mathbf{x}, t)$, which is a field of white noise that appears in the modified Schrödinger equation (a stochastic differential equation). In (2.9) these white noise processes become Brownian motion processes, $B(\mathbf{x}, t)$, because of the general fact that

$$\int_0^t w(\tau)d\tau = \int_0^t dB(\tau) = B(t) - B(0)$$

and we set $B(0) = 0$ by convention.

7 It can indeed be shown that (2.14) is a legitimate probability measure. See, for example, Ghirardi, Pearle, and Rimini (1990).

8 The discussion can just as well be given in terms of density operators, but it is less intuitive. See Pearle (1989) and Ghirardi, Pearle, and Rimini (1990).

9 Ghirardi, Pearle, and Rimini (1990) provide the rigorous discussion.

10 A stochastic process, $X(t)$, is a martingale if: (i) $E[|X(t)|] < \infty$ for all t, and (ii) for any $t_1 < \cdots < t_n < t, E[X(t)|X(t_1) = x_1, \ldots, X(t_n) = x_n] = x_n$, where $E[\,\cdot\,]$ is an expectation. Condition (ii) is called the 'martingale property', and martingales are sometimes called 'fair game processes', because, intuitively, the expectation value of one's 'points' is constant in time (letting $X(t)$ be one's points at time t). Note, however, that the martingale property does not imply that such games have no end, i.e., that no player's points will go to zero.

11 In so doing, they follow a suggestion of Bell's (1990), and ultimately of Schrödinger's (1952). Bell's suggestion could be made more precise by considering density of mass and charge, but I shall not speculate on details here.

12 See, for example, Pearle (1992b).

13 On the other hand, in a talk in Bielefeld in 1994, David Albert pointed out that if a boundary exists, we could never find it. The idea is basically that if there is a

fundamental limit to the precision with which we can discern distances, then, in the context of CSL, it would be necessary to exceed this limit in order to discover exactly what the limit is. Albert therefore saw no real problem to postulating such a fundamental limit, for it would, in a sense, be able to explain why vagueness seems to us to be characterized by fuzzy boundaries.

14 See, for example, Albert and Vaidman (1989a, 1989b).
15 One should perhaps keep in mind that the examples involve photons, whereas CSL was not developed explicitly to account for photons. However, I believe that the general points raised by the examples are valid; hence I ignore this issue for convenience.
16 See Aicardi *et al.* (1991).
17 See, for example, Albert (1990).
18 See, for example, Stapp (1992).

Chapter 3

1 The only two proponents of the bare theory that I know of are Coleman (1994) and Albert (1992), and Albert maybe should not be considered an advocate. I am indebted to the talk given by Jeff Barrett at the University of Utrecht (June, 1996), and to discussion with Jeff Barrett about the bare theory.
2 Bub, Clifton, and Monton (1997) make this same point.
3 Everett (1957, p. 456).
4 See, for example, Saunders (1993).
5 There are, in fact, many versions of the many minds interpretation. See, for example, Albert (1992), Albert and Loewer (1988), and Lockwood (1996).
6 Griffiths' original paper is his (1984). Since then, there have been many papers on the consistent histories approach and related approaches; for just a few examples, see Omnés (1989, 1994) and Gell-Mann and Hartle (1990, 1993), and Zurek (1993).
7 There are in the literature other suggestions for the definition, but the details of the differences among different definitions are not important here. See Gell-Mann and Hartle (1990).
8 The proof appears in Griffiths (1984).
9 Griffiths uses this language in, for example, his (1993), where he says that the consistent histories approach is a 'realistic interpretation of quantum mechanics' (p. 1602).
10 Griffiths (1993, p. 1605)
11 Indeed, Kent (1995) has shown that the histories chosen by the Kochen–Dieks–Healey modal interpretation cannot be consistent. Yet, as shown in the next chapter, this modal interpretation has a completely satisfactory (and, apparently, classical) dynamics. So why is consistency essential?
12 Some of the discussion to follow in the text owes a great deal to Butterfield (1996a, 1996b).
13 To be more precise, if the distribution does not assign non-zero probability to just a countable subset of the continuously many elements, then each has probability 0.
14 One reply to this argument that is sometimes made is that although any density (that assigns non-zero probability to more than a countable subset) will assign equal probability (namely, 0) to all elements in the set, it will not assign equal probability to subsets of worlds that are 'equal in size'. This reply is obviously flawed, because prior to the definition of some measure over the set, there is no sense in which two subsets of worlds can be said to be 'of equal size'.
15 This point has been made many times. See, for example, Lockwood (1996).

Chapter 4

1 See, for example, Friedman and Putnam (1978).
2 See Putnam (1969)

3 See Putnam (1974).

4 See Kochen and Specker (1967) for the original paper. Clifton (1993), Mermin (1990, 1994), and Peres (1994) provide good contemporary expositions.

5 This analogy was first given, as far as I know, by Putnam (1969).

6 For further critique of the quantum logic interpretation, see, for example, Dummett (1976) and Redhead (1987).

7 It first appears in Friedman and Putnam (1978).

8 When they give their definition of conditional equivalence, Friedman and Putnam do not mention explicitly that the definition is sensible only when P and Q (the conditionally equivalent propositions) are compatible. Indeed, some incompatible propositions are equivalent conditional on $|\psi\rangle$, and yet $|\psi\rangle$ assigns them different probabilities.

9 I am indebted for Michael Friedman for suggesting to me something like the idea that follows in the text. Nonetheless, nothing that I say here should be foisted upon him.

10 'Faux-Boolean algebra' is an odd name, I agree. The name is meant to convey the fact that these algebras are not in general Boolean, but that, given the right sort of quantum probability measure over them, they cannot be discerned to be non-Boolean. The perhaps more natural name 'pseudo-Boolean algebras' has already been taken for a different sort of structure (sometimes called a Heyting algebra) — I am grateful to John Bell and Rob Clifton for pointing this fact out to me.

11 There are more abstract ways to define the same type of structure — ways that do not make reference to Hilbert space. However, the definition given here is sufficient for my purposes. Perhaps the best example of the more abstract approach is Zimba and Clifton (1997), but see also Bell and Clifton (1995).

12 The proof of theorem 4.1 is in Dickson (1995b).

13 The proof of this theorem, and of the next, appears in Dickson (1995c). The proof of this theorem relies a point made originally (as far as I know) by Cufaro-Petroni (1992), and more importantly, on a theorem by Halmos (1969).

14 See van Fraassen (1979, 1981, 1990) for earlier expositions. The most recent exposition is in chapter 9 of his (1991), which is a revision of his (1981).

15 Van Fraassen uses the terms 'value state' and 'dynamic state', but I stick with the terms that I have already used for uniformity.

16 For details, see van Fraassen (1991, pp. 222–224). Elsewhere in his (1991), van Fraassen discusses other types of measurement, though as far as I can tell, he does not anywhere give a complete account of all types of measurement (nor does he claim to).

17 See van Fraassen (1991, p. 287), though he does not state the rule in exactly the same terms as used here.

18 See Dickson (1995a) for a more extended argument to this end.

19 See, for example, van Fraassen (1991, ch. 7)

20 Indeed, Healey's (1989) modal interpretation is quite complex. Healey's conditions on $\mathscr{A}_W$ make it difficult to say what the structure of the algebra of definite-valued events is without going into a great deal of detail. I shall not do so here.

21 The formulation to follow is due to Clifton (1995a) and Vermaas and Dieks (1995).

22 See Clifton (1995a, 1995b). This result holds for Healey only prior to the application of his other principles — see note 21. Vermaas and Dieks (1995) pointed this result out independently of Clifton.

23 A proof of this theorem is in Schrödinger (1935b). (A version of the proof in modern notation, but still based on Schrödinger's proof, can be found, for example, in Dickson (1994a). Kochen bases his modal interpretation not on the biorthogonal decomposition theorem, but on the polar decomposition theorem. See Kochen (1985) for a statement of the theorem, and Reed and Simon (1979, pp. 197–198) for a proof.)

24 Arntzenius (1990a).

25 See Vermaas and Dieks (1995).

26 See Clifton (1995a, 1995b).

27 Clifton (1995a).

28 Clifton (1995a).

29 See Dickson (1995b) for the proof.

30 See Albert (1990, 1991), Albert and Loewer (1993) and Elby (1993, 1994) for various statements of what is essentially the same problem.

31 See, for example, Bacciagaluppi and Hemmo (1994, 1997a) and Dickson (1994a).

32 See Bacciagaluppi (1997).

33 Bacciagaluppi (1997).

34 See, e.g., Bub (1995); cf. Bub and Clifton (1996).

35 Bub and Clifton (1996) prove a similar theorem, based on different conditions, but theirs also applies only to pure states.

36 As far as I know, the phrase 'naïve realism about operators' is due to Goldstein (1996a).

37 Dieks (1997) seems to have abandoned this view. Instead, he says that we simply choose a factorization that is convenient, or appropriate, for the problem at hand.

38 See Dieks (1988a, 1989) and Vermaas and Dieks (1995).

39 Bacciagaluppi and Dickson are motivated by dynamical considerations to propose this approach. For a more detailed discussion, see Bacciagaluppi and Dickson (1997).

40 The original proof is by Bacciagaluppi (1995). See also Bacciagaluppi (1996) for further discussion. As I mentioned in note 37, Dieks' response to this situation is to adopt a kind of 'interest-relative' preferred factorization.

41 See Vermaas (1996a) for a limited definition of such joint probabilities. See Vermaas (1996b) for an argument that no satisfactory general expression is to be had.

42 Dieks (1988a, 1989), Vermaas and Dieks (1995).

43 Nick Reeder (personal communication) first pointed this fact out to me.

44 Cf. Fevrier (1937), Reichenbach (1944), and Putnam (1957), but note that the proposal here is certainly not what they had in mind.

45 See Dickson (1996b).

46 Van Fraassen (1991, ch. 9) says something along these lines, though he does not consider faux-Boolean algebras. The definition of the modal operators given in the text is also motivated by van Fraassen.

47 This relation of accessibility is an equivalence relation — we are dealing with the modal system S5. However, clearly other accessibility relations could be defined, thereby yielding other systems of modal logic.

48 Indeed, Vink (1993) has shown that by choosing Bub's preferred observable to be a coarse-grained position observable, then taking the limit as the observable's spectrum becomes continuous, one ends up with Bohm's theory (given a suitable choice for a stochastic dynamics on the original coarse-grained observable), which is completely deterministic.

49 For a long and far more detailed account of dynamics for modal interpretations, see Bacciagaluppi and Dickson (1997). See also Dickson (1997) and Bacciagaluppi (1996) for discussion.

50 As I said earlier, this entire discussion takes place in the context of finite-dimensional Hilbert spaces. Many of the arguments carry over to infinite dimensions, but some do not. The interested reader is invited to consult the papers mentioned throughout this chapter for further details.

51 The proof is in Bacciagaluppi, Donald, and Vermaas (1996). However, as they point out, the evolution can be quite erratic (though still continuous). Moreover, the result is less firm in the case of infinite-dimensional Hilbert spaces. There, under certain circumstances, discontinuities are possible.

52 Vink (1993).

53 A more detailed argument is in Dickson (1997), though even there, the argument is sketchy. Nonetheless, I am convinced that no generally compelling conditions will select a unique dynamics for modal interpretations. Dickson (1997) discusses this point, and in particular how one might react to it.

54 Bell's choice is in his (1987b), and is motivated by Bohm's (1952) theory, in which the transition probabilities are all 0 or 1, and the velocity of a particle (which determines

which transitions probabilities are 1) is given by the current divided by the probability density. See chapter 5.

Also, Bell's expression for $T_{j|i}$ is not exactly (4.24), but

$$T_{j|i} := \begin{cases} \dfrac{J_{j|i}}{p_i} & \text{for } J_{j|i} > 0 \\ 0 & \text{for } J_{ji} \leq 0. \end{cases}$$

The difference, however, is irrelevant, at least if p_i has only isolated zeros. In fact, the two choices can differ only at points with $p_i = J_{j|i} = 0$, where (4.24) may be infinite, while Bell's expression is zero. The choice (4.24) is continuous at these exceptional points.

55 For the details of how to implement this trick (which came from James Cushing (personal communication)), see Bacciagaluppi and Dickson (1997).

56 Work by Bacciagaluppi and Hemmo (1997b) suggests that in any case the physical state of a quantum system in the Kochen–Dieks–Healey interpretation has nearly nothing to do with the observed outcomes of measurements on it.

57 See Vink (1993) for the proof.

58 One cannot say with confidence, however, that the notion of a faux-Boolean algebra, or some reasonably similar structure, can be defined in this case, because of course the 'eigenspaces' of the position observable are, in fact, not in the Hilbert space, which is why they are more properly called 'improper eigenspaces'.

Chapter 5

1 The original paper is Bohm (1952). For some later developments, see Holland (1993) and Bohm and Hiley (1993).

2 Such a formulation was given by, for example, Bell (1987d) and Dürr, Goldstein, and Zanghì (1992).

3 The main paper along these lines is Dürr, Goldstein, and Zanghì (1992), but there are several other papers developing different aspects of this formulation of Bohm's theory. For a review of the approach, see Dürr, Goldstein, and Zanghì (1996).

4 Dürr, Goldstein, and Zanghì sometimes give the impression that they have derived Bohmian mechanics *a priori* (from more or less necessary symmetry conditions). However, their argument relies at one point on an argument from simplicity — i.e., by the end of the story one can only say that (5.9) is the 'simplest' expression meeting certain symmetry conditions. Of course, it is notoriously difficult to justify any such claim rigorously, the problem being that we do not have a well-defined notion of simplicity.

5 See Brown and Anandan (1995).

6 Holland (1993) contains this view, and his arguments for it, in great detail.

7 Bohm and Hiley (1993, p. 31).

8 Bohm and Hiley (1993, p. 31).

9 Bohm and Hiley (1993, p. 32).

10 Bohm and Hiley (1993, p. 37).

11 This argument appears, for example, at Holland (1993, p. 78).

12 Holland (1993, p. 226)

13 Holland (1993, p. 421).

14 Recall Albert's objection to CSL, discussed in chapter 2.

15 The argument appears, for example, in Dürr, Goldstein, and Zanghì (1992).

16 The statement that follows in the text is a trivial consequence of theorem 12.2.2 in Beltrametti and Cassinelli (1981).

17 See, for example, Bell (1987d).

18 The point is essentially the point that Albert made against CSL and that I attempted to answer — see chapter 2.

19 For arguments specific to Bohm's theory, see, for example, Bohm and Hiley (1993, ch. 8). For a general discussion of decoherence, see, for example, Zurek (1993).

20 Ehrenfest's theorem is discussed in most textbooks on quantum mechanics — see, for example, Cohen-Tannoudji *et al.* (1977). It is often mistakenly supposed to solve the problem of the classical limit even in standard quantum mechanics.

21 See Dürr, Goldstein, and Zanghi (1992) and Valentini (1991a, 1991b).

22 Valentini (1991a) has a similar proof.

23 Dürr *et al.* make this argument in several places, but again, the main reference is their (1992).

24 See Valentini (1991a, 1991b).

Chapter 6

1 Bell's original paper is his (1964), reprinted in his (1987e). He adapted the argument to a probabilistic setting in his (1987c).

2 The analysis discussed here in the text was anticipated in Clauser and Horne (1974), Suppes and Zanotti (1976, pp. 445–455), and van Fraassen (1982), but was first fully and clearly articulated by Jarrett (1984). See also Jarrett (1989, pp. 69–70), Ballentine and Jarrett (1987), and Shimony (1986, pp. 182–203).

3 The conditions as given in the text are not exactly those given by Jarrett, who considers not only the complete state of the pair of particles, but also the complete states of the apparatuses. Neither are the terms his; they are from Shimony (1986), who (intentionally) *does not* consider the complete state of the apparatuses.

4 For related arguments along these general lines, see Jones and Clifton (1993), Cushing (1994a), and Dickson and Clifton (1997).

5 The proof of Maudlin's claim (which appears in his (1994)) is not given in his book, and proceeds along lines slightly different from the proof of Jarrett's result, but it still uses only standard probability theory.

6 In the end, I do not find this 'holistic' view at all plausible, however, and certainly it is *not* entailed by standard quantum mechanics. After all, standard quantum mechanics *does* assign to each particle a definite (albeit mixed) state, and we can just as well view the correlations between them (more precisely, the systems that have them) as the result of causal influences.

7 The model in question was devised by Jones and Clifton (1993). This model is important because it shows that Jarrett's two conditions — which appear to be independent — can be rendered *non*-independent by the details of a model.

8 Curiel (1996) discusses this point in detail.

9 See also Bell (1987c), Clauser and Horne (1974), and the extensive review article by Clauser and Shimony (1978). Since Bell's original theorem, perhaps the most important development in this area is the theorem by Greenberger, Horne, and Zeilinger (1989) (cf. Greenberger, Horne, Shimony, and Zeilinger (1990) and Clifton, Redhead, and Butterfield (1991)), who prove a similar result, but without recourse to inequalities.

10 Moreover, though this fact does not concern me here, Bell's inequality is violated by experiment, but only given some assumptions. For example, one assumes implicitly in derivations of Bell's inequality that the particles leave the source in exactly opposite directions, but this assumption is an idealization, and there are problems with more general versions of Bell's inequality that attempt to take this fact into account. See Peres (1978). Marshall, Santos, and Selleri (1983) give a critical review of the assumptions required by the tests performed by Aspect and coworkers (1981, 1982a, 1982b), the most famous to date. Also, there are more general epistemological objections to the effect that experimental results can often or always be 'reinterpreted'. See, e.g., Krips (1987, pp. 163–164).

11 That Suppes and Zanotti's derivation (which appears in their (1976) should go through using factorizability is not at all surprising, and what follows in the text is little more than a transcription of the proof of Suppes and Zanotti's result, as given by van Fraassen (1982).

12 That is, if we denote the set of λ for which (6.14) fails by 'Λ_{fail}' then

$$\int_{\Lambda_{\text{fail}}} \rho(\lambda)\, d\lambda = 0.$$

13 A boring and serpentine formal proof of the result appears in Dickson (1996a), whose proof relies on a theorem by Pagonis, Redhead, and Clifton (1991).
14 See Fine (1982a, 1982b).
15 Fine's original results were non-dynamical, but they can easily be extended to cover the case being considered here (where the complete states are given at a time prior to measurement). I show how to do so, and how to generalize Fine's result to N particles, in Dickson (1996a).
16 Fine (1982a, p. 294).
17 The argument as given by these authors — for example, Svetlichny, Redhead, Brown, and Butterfield (1988) and Butterfield (1992a) — is in terms of the existence of joint probabilities for non-commuting observables, but their points are easily translated into the terms used here.
18 Butterfield (1992a, pp. 77–79).
19 Butterfield (1992a, 1995).
20 As Goldstein (1996b) reports, Wigner (1983, p. 53) says that Bell's theorem is the 'truly telling argument against' Bohm's theory. And more recently, Gell-Mann (1994, pp. 170–172) too suggested that Bell's theorem puts Bohm's theory to rest.

Chapter 7

1 To represent the state as a stochastic process is, strictly speaking, to assume that the λ are scalars. Of course, most likely they would not be scalars, and in that case one would have to be more careful about how to define the mathematical object $L(t)$. However, there is no serious problem here, as far as I can tell. There are, for example, adequate definitions of 'stochastic processes' as time-indexed families of random vectors, or even random matrices, rather than random variables. Nothing in my discussion in the text relies essentially on the assumption that λ is a scalar.
2 That is, $\rho(\lambda_n | \lambda_0)$ is defined so that for any subset, $M \subseteq \Lambda$, and any time t^n

$$p\left(L(t^n) \in M \,\Big|\, L(t_0) = \lambda_0\right) = \int_M \rho(\lambda_{t^n} | \lambda_0)\, d\lambda_{t^n}.$$

3 Or, one could interpret the probabilities $p^{\lambda_0}(d_n^t)$ as conditionals with probabilistic consequents: if the outcome is to be at t^n, then the probability that it will be d_n^t is $p^{\lambda_0}(d_n^t)$. This interpretation would require slight modifications in a few things that I say.
4 That is, $p^{\lambda_n}(d_n^t) = \chi_{\Lambda_{d_n}}(\lambda_n)$, which is 0 for all $\lambda_n \notin \Lambda_{d_n}$ and 1 for all $\lambda_n \in \Lambda_{d_n}$.
5 A similar point was made by Jarrett (1986).
6 Born (1971, pp. 170–171).
7 Whether Einstein distinguished these concepts is, I think, an open question. Howard (1985, 1989) argues that he did.
8 Finding appropriate language is difficult here. By 'a part' one cannot mean a part in the usual sense, i.e., a part that can be removed, or thought of as a object unto itself, occupying its own continuously connected region of space-time.
9 Healey (1991, 1994) and Teller (1989) are two proponents of holism, though the idea can be found elsewhere as well.
10 The quotation marks are to avoid introducing the technical jargon needed for speaking about holistic objects — the reader who is concerned can probably make the appropriate translations, or see Healey (1991) for a careful discussion.
11 Howard argues that separability — and he appears to mean the same thing by it as do I (Howard 1989, pp. 226–227) — is connected with outcome independence. However,

partly for reasons given by Maudlin (1994, pp. 97–98), and partly because I think separability is a reasonable condition, while outcome independence is not, I think Howard is wrong to make this connection. See also the discussion by Laudisa (1995), who gives additional reasons for disliking holism.

12 This assumption can be weakened considerably. First, note that the record need not be of the traditional sort — there merely needs to be some reliable 'trace' of the result in the forward light cone of the outcome-event. Second, this trace need not be completely reliable. That is, all one really needs is a reliable correlation between the outcome-event and some later events in its forward light cone.

13 Well, there is perhaps one way to get around this particular argument. It could be that the *time* of measurement for both particles is fixed, so that one particle *does* 'know' when the other is measured. However, it is unclear in this case what it would mean for the particles to 'decide' on results at times other than the time at which they are measured.

Chapter 8

1 See Maudlin (1994).

2 Putnam's paper is his (1967) — the paper was first read to the American Physical Society in 1966, and Rietdjik's is his (1966). Rietdjik made the argument again in his (1976). Maxwell has also made this argument in his (1985) and (1988).

3 Stein (1970, 1991).

4 Putnam (1967, p. 240).

5 Putnam (1967, p. 247).

6 Rietdjik (1966, 1976).

7 Rietdjik (1966, p. 342).

8 Maxwell (1985, p. 25).

9 Dieks (1988a) also made an attempt to formulate a notion of probabilism that escapes the block-universe argument, but, as Stein (1991, p. 152) has pointed out, Dieks' attempt appears to presuppose a notion of absolute simultaneity.

10 Stein (1970, 1991).

11 Rietdjik (1976) and Maxwell (1985, 1988).

12 Putnam (1967, p. 246).

13 Maxwell (1985, pp. 27–28).

14 Maxwell (1985, p. 28).

15 See, for example, Reichenbach (1956). However, note that Reichenbach in fact defined screening-off in terms of four independent conditions, of which the one given in the text is the first.

16 See, for example, Redhead (1987) for a more detailed discussion.

17 Maudlin (1994).

18 For a discussion, see Arntzenius (1990b).

19 Curiel (1996) has made this point in an extended essay. He argues that even in the context of a quite specific model, namely, one given by Jones and Clifton (1993), it is a difficult matter to decide how the causes are acting.

Chapter 9

1 Maudlin (1994, p. 199)

2 For an example of such a theory, see Hellwig and Crause (1970), but also the criticism by Aharonov and Albert (1981).

3 See, for example, Butterfield (1992a, 1992b), who argues that the violation of locality in the standard interpretation is causal — i.e., events in one region of space-time causally influence events in some other space-like separated region.

4 See Stapp (1992) for the former suggestion, and Ghirardi, Grassi, and Pearle (1990a) and Penrose (1996) for the latter. One side effect of this interpretation would be to rule out

an advantage that Albert (1994) finds in CSL, namely, its ability to underwrite statistical mechanics, and especially to render otherwise unacceptable explanations in statistical mechanics acceptable. Albert's argument works (as he acknowledges) only if the probabilities in CSL are genuine, irreducible, physical chances. However, these interpretations of CSL make them merely epistemic.

5 Strictly speaking, it does not. Or rather, the equivalence holds only in the limit of infinite time. However, there is really no problem here. Although I assumed absolutely strict correlations in the arguments of chapters 6 and 7, small deviations from them would weaken the conclusions to 'near determinism', which is still violated by CSL, in which probabilities are, in general, not even near 0 or 1.

6 See Ghirardi, Grassi, and Pearle (1990b), Ghirardi and Pearle (1990), and Pearle (1990, 1992a). A problem that plagued these theories was the production of infinite energy, due essentially to the fact that white noise has no upper bound on its frequency. Pearle (private communication) has found that with more realistic stochastic processes, in which there is an upper bound on the frequency, this problem is resolved.

7 I have used the word 'appears' in summarizing the problem because it is not obvious to me that there *is* a problem. Whether retrodiction is possible in a given stochastic theory is a fascinating, but difficult, question. The advocates of CSL seem to think that the answer is 'no' — at least for CSL — and I follow them here, but not with complete confidence.

8 Ghirardi and Pearle (1990).

9 Ghirardi and Pearle (1990, p. 45). Essentially the same statement is made by Ghirardi, Grassi, and Pearle (1990b). An apparently different proposal is made by Ghirardi and Grassi (1994) — see note 11.

10 Ghirardi and Pearle (1990, p. 45).

11 This claim is a consequence of the no-signalling theorem for CSL. See Butterfield *et al.* (1993a, 1993b) for a proof and discussion.

12 See, for example, Dieks (1994) and Healey (1989, chs.4, 5; 1997)

13 One easy way to secure this assumption is to prepare a system in the singlet state, then 'measure' whether it is in the singlet state by passing it through a filter that allows only systems in the singlet state to pass. (Of course, practical design of such a filter is no simple task.)

14 Or, the singlet state plus some theoretical state. However, we may assume that the pair already began in the singlet state (see note 14), so that in this case the physical and theoretical states are the same.

15 Hardy (1992).

16 The reason is twofold. First, slight inhomogeneities in the interaction (for example, due to unavoidable background magnetic fields) turn the exact degeneracy of the ideal interaction into 'near' degeneracy. Second, it has been shown that in such cases, decoherence makes the reduced state of the apparatus diagonal in a basis that is nearly that picked out by the pointer-observable. See Bacciagaluppi and Hemmo (1994, 1997a) and Dickson (1994a).

17 More generally, the joint probabilities for properties in different subsystems will lead to a violation of independent evolution. As I discussed briefly in chapter 4, the joint probabilities put restrictions on the evolution for the subsystems, and in general the effects of these restrictions do not disappear entirely, even when we sum over the other subsystem. There does not seem to be any way to avoid the dependence of the evolution of one subsystem on the physical state of another.

18 Three places where Healey discusses non-locality in great detail are his (1989, 1991, 1994).

19 Dieks (1994, p. 2297).

20 One could postulate a dynamics in which independent evolution failed even in these cases, but, as I discussed in chapter 4, there exist dynamics, such as the one given there, in which it does not.

21 Dieks (1994, p. 2297).

22 Specifically Healey finds that the following condition is violated: 'if spacetime regions R_e, R_f are space-like separated from one another, then no process can directly connect an

event *e* within R_e to an event *f* within R_f' (Healey 1994, p. 39). However, he argues that the sort of violation of this condition found in his modal interpretation does not strictly violate special relativity, because there is no 'causal order' of space-like separated events, and therefore no causal paradox is generated. (However, whether one buys this argument or not, no-signalling theorems already appear to guarantee no *empirical* conflict with special relativity.)

23 Dieks (1994).

24 See Fleming (1989) and Fleming and Bennett (1989). Fleming advocates the projection postulate (relativized to a hyperplane) as well, but his central metaphysical claim can be made without reference to the projection postulate.

25 Healey (1994, p. 39).

26 The proof appears in Dickson and Clifton (1997).

27 The preliminary analysis occurs in Dickson and Clifton (1997). They show that in certain cases, at least (those in which the calculation of the determinate sublattices for the subsystems of interest is not too difficult), Bub's interpretation is Lorentz-invariant only if *R* is the identity (in which case the measurement problem is apparently not solved) or if one draws a fundamental distinction between measured systems and apparatuses (by choosing $R = 1$ for measured systems, but non-trivial *R* for apparatuses), in which case the measurement problem is solved, but not plausibly.

28 I say 'technicalities' because, although Earman (1986) has shown that Newtonian mechanics is not strictly deterministic — there exist strange scenarios in which determinism fails due, for example, to infinite velocities — the cases where determinism fails are not physically interesting, as far as I can tell.

29 The point here is similar to one made by Earman (1986). If particles could enter the light cone 'unannounced', then the state $L(t)$ for some $t > t_0$ would *not* be fixed by the earlier state, $L(t_0)$ (plus the Hamiltonian).

30 There is one complication. Englert *et al.* (1992) have shown that in Bohm's theory, there are times when the particle will be at one detector, but the *other* detector fires. (See also Dewdney, Holland, and Squires (1993) and Dürr *et al.* (1993).) However, it still is true that the outcome — the flashing of a detector at some point — must in principle be determined by the initial state.

31 Bell (1987d). When the particles are in the singlet state, the expression is:

$$v_\alpha = g^\alpha(t) \frac{|\varphi_\alpha^-|^2 |\varphi_\beta^+|^2 - |\varphi_\alpha^+|^2 |\varphi_\beta^-|^2}{|\varphi_\alpha^-|^2 |\varphi_\beta^+|^2 + |\varphi_\alpha^+|^2 |\varphi_\beta^-|^2},$$

where $g_\alpha(t)$ characterizes the coupling between the spin of α and its apparatus, and $\varphi_\alpha^\pm$ is a time-dependent wave function, a narrow wave packet moving towards the $\pm$ detector for α. (Bell's expression assumes an infinite mass for the particles, which allows one to ignore the contribution to the current from the free Hamiltonian for each particle. Although this assumption is an idealization, the essential point here does not rely on it and could be made also in the general case.)

32 For an extensive account of the EPR–Bohm experiment in Bohm's theory, including an account of the trajectories of the particles, see Dewdney, Holland, and Kyprianidis (1986, 1987).

33 This case is essentially the one envisaged by Maudlin (1994, pp. 134–135).

34 Indeed, it *does* fail in some modified versions of Bohm's theory, suggested by Bohm and Vigier (1954) and more recently advocated by Bohm and Hiley (1993). In those theories, it is suggested that small random fluctuations occur in the trajectories of the particles. The hope is apparently to recover more naturally the use of $|\psi|^2$ as a probability. However, it is not clear that the hope has been fulfilled by these stochastic versions of Bohm's theory. The results derived within those theories seem no better than the results of Valentini, discussed in chapter 5. (Nor, perhaps, are they any worse. They share with Valentini's results the lack of a minimum *time* in which the probability distribution is driven to $|\psi|^2$. Hence it is compatible with these results that the present probability

distribution is nothing close to $|\psi|^2$.) Moreover, the analysis here shows that such a modified Bohm's theory *must* put up with forms of non-locality not needed in the deterministic version of Bohm's theory. Indeed, for this point, I can rely in part on the analysis of Bohm and Hiley (1989) themselves. They recognize that there exists instantaneous action at a distance in their theory. Moreover, they seem to have concluded on the basis of technical difficulties that the stochastic fluctuations cannot be made covariant at the level of individual fluctuations.

35 I deviate from Bell's notation, in which $C(i, j)$ is written $P(i, j)$, in order to avoid introducing yet another meaning for 'P'.

36 Lewis (1986).

37 Lewis (1973).

38 Lewis is known for advocating the non-backtracking method, but my advocacy of the backtracking method is not inconsistent with Lewis' views. Lewis is concerned in the first place with the evaluation of everyday counterfactuals, in the context of everyday concerns. He admits, however, that given a special set of concerns, one might have different criteria for the evaluation of counterfactuals. In our case, we have the special concern of discovering the features of a scientific theory, and it seems reasonable in this case to place high priority on not violating that theory.

39 Lewis (1973).

40 See, for example, Salmon (1984).

41 Erik Lindlin brought this objection to my attention.

42 Hardy (1992); see also Hardy and Squires (1992).

43 Hardy's discussion, and mine here, is within particle mechanics, not field theory. At present, Bohmian field theories are manifestly non-covariant, and must, therefore, adopt a preferred frame. (See Bohm, Hiley, and Kaloyerou (1987).)

44 Berndl and Goldstein (1994). In his reply, Hardy (1994) agrees with their argument. See also the comments by Clifton and Niemann (1992) and Dickson and Clifton (1997).

45 This fact is one of the important lessons from Maudlin's (1994) book.

46 Various special cases have been discussed, though they do not show any promise of leading to a general Bohmian interpretation of the Klein–Gordon equation. See, for example, Cufaro-Petroni *et al.* (1984) and references therein; and see Dewdney *et al.* (1992, pp. 1259ff) for a general discussion. One important difficulty is the same one that led to the initial rejection of the Klein–Gordon equation, namely, the fact that the time-component of the 4-current (i.e., the density) is not positive-definite. See Holland (1993, pp. 498–502) for a discussion. Another supposed difficulty is the fact that the x that appears in the Klein–Gordon 4-current is not the eigenvalue for any operator. Thus it is concluded that x cannot be taken as a 'position' (because there is no Hermitian operator, X, such that $X|x\rangle = x|x\rangle$). See Prugovečki (1984, pp. 71–83). However, this argument is flawed. Bohm's theory need not subscribe to the view that position must be representable as an operator. Indeed, Bohm's theory is an *interpretation* of the quantum formalism, and therefore can interpret that formalism in such a way that the x appearing in the current, or the density, *does* refer to a physical 'position'. See also the discussion in the subsection 9.4.2.3.

47 What follows in the text is certainly nothing new. Bohm (1953) was the first to consider a Bohmian interpretation of the Dirac equation. An interpretation within the 'stochastic' modification of Bohm's theory appears in Bohm and Hiley (1989). See Holland (1992), or pp. 503–518 of his (1993), for a detailed treatment, including an interesting discussion of the Klein paradox and *zitterbewegung*.

48 Holland (1993, pp. 503–509).

49 Holland's expression for the 4-velocity is:

$$u^\mu = \frac{j^\mu}{\sqrt{(\psi^\dagger \gamma^0 \psi)^2 + (\psi^\dagger \gamma^1 \gamma^2 \gamma^3 \psi)^2}} = \frac{j^\mu}{\sqrt{j^\mu j_\mu}}$$

(summing over repeated indices).

50 See, for example, Sakurai (1967, p. 101).
51 See, for example, Sakurai (1967, p. 102).
52 Holland (1992, p. 1298).
53 Indeed, there does exist a covariant 'Bohm-like' theory, that of Mackman and Squires (1995), which is a version of Bohm's theory with a retarded quantum potential. They show explicitly how their theory avoids Hardy's argument. However, unlike Bohm's theory, their theory is not empirically equivalent to standard quantum mechanics. In addition, Berndl, Dürr, Goldstein, and Zanghì (1996) have argued that a covariant Bohm-like theory is not impossible.
54 Malament (1996). His argument was, however, directed in the first instance towards Fleming's (1989) hyperplane dependence.

References

Aharonov, Y, and Albert, D. (1981). Can We Make Sense of the Measurement
     Process in Relativistic Quantum Mechanics?, *Physical Review Letters* **24**,
     359–370.

Aicardi, F., Borsellino, A., Ghirardi, G. C., and Grassi, R. (1991). Dynamical
     Models for State-Vector Reduction: Do They Ensure that Measurements Have
     Outcomes?, *Foundations of Physics Letters* **4**, 109–128.

Albert, D. (1990). On the Collapse of the Wave-Function, pp. 153–165 in *Sixty-Two
     Years of Uncertainty*, ed. A. Miller (Plenum, New York).

Albert, D. (1991). Some Alleged Solutions to the Measurement Problem, *Synthese*
     88, 87–98.

Albert, D. (1992). *Quantum Mechanics and Experience* (Harvard University Press,
     Cambridge, MA).

Albert, D. (1994). The Foundations of Quantum Mechanics and the Approach to
     Thermodynamics Equilibrium, *British Journal for the Philosophy of Science* **45**,
     669–677.

Albert, D. and Loewer, B. (1988). Interpreting the Many-Worlds Interpretation,
     Synthese **77**, 195–213.

Albert, D., and Loewer, B. (1993). Non-Ideal Measurements, *Foundations of
     Physics Letters* **6**, 297–305.

Albert, D., and Vaidman, L. (1989a). On a Proposed Postulate of State-Reduction,
     Physics Letters A **139**, 1–4.

Albert, D., and Vaidman, L. (1989b). On a Theory of the Collapse of the
     Wave-Function, pp. 7–16 in *Bell's Theorem, Quantum Theory, and Conceptions
     of the Universe*, ed. M. Kafatos (Kluwer, Dordrecht).

Arntzenius, F. (1990a). Kochen's Interpretation of Quantum Mechanics,
     pp. 241–249 in *PSA 1990* vol. I, eds. A. Fine, M. Forbes, and L. Wessels, L.
     (Philosophy of Science Association, East Lansing, MI).

Arntzenius, F. (1990b). Causal Paradoxes in Special Relativity, *British Journal for
     the Philosophy of Science* **41**, 223–243.

Aspect, A., Grangier, P., and Roger, G. (1981). Experimental Tests of Realistic
     Local Theories via Bell's Theorem, *Physical Review Letters* **47**, 725–729.

Aspect, A., Grangier, P., and Roger, G. (1982a). Experimental Realization of
     Einstein-Podolsky-Rosen-Bohm *Gedankenexperiment*, A New Violation of
     Bell's Inequalities, *Physical Review Letters* **48**, 91–94.

Aspect, A., Dalibard, J., and Roger, G. (1982b). Experimental Tests of Bell's

231

Inequalities Using Time-Varying Analyzers, *Physical Review Letters* **49**, 1804–1807.

Bacciagaluppi, G. (1995). A Kochen-Specker Theorem in the Modal Interpretation of Quantum Mechanics, *International Journal of Theoretical Physics* **34**, 1206–1215.

Bacciagaluppi, G. (1996). Topics in the Modal Interpretation of Quantum Mechanics, Ph. D. Dissertation (Cambridge University).

Bacciagaluppi, G. (1997). Delocalised Properties in the Modal Interpretation of a Continuous Model of Decoherence, unpublished manuscript.

Bacciagaluppi, G., and Dickson, M. (1997). Modal Interpretations with Dynamics, University of Cambridge preprint.

Bacciagaluppi, G., Donald, M., and Vermaas, P. (1996). Continuity and Discontinuity of Definite Properties in the Modal Interpretation, *Helvetica Physics Acta* **68**, 679–704.

Bacciagaluppi, G., and Hemmo, M. (1994). Making Sense of Approximate Decoherence, pp. 345–354 in *PSA 1994* vol. I, eds. D. Hull, M. Forbes, R. and Burian (Philosophy of Science Association, East Lansing, MI).

Bacciagaluppi, G., and Hemmo, M. (1997a). Modal Interpretations of Imperfect Measurements, *Studies in History and Philosophy of Modern Physics*, forthcoming.

Bacciagaluppi, G., and Hemmo, M. (1997b). State Preparation in the Modal Interpretation, forthcoming in *Minnesota Studies in Philosophy of Science*, eds. R. Healey and G. Hellman.

Ballentine, L., and Jarrett, J. (1987). Bell's Theorem: Does Quantum Mechanics Contradict Relativity?, *American Journal of Physics* **55**, 696–701.

Bell, J. (1964). On the Einstein–Podolsky–Rosen Paradox, *Physics* **1**, 195–200.

Bell, J. (1987a). Are There Quantum Jumps?, pp. 201–212 in *Speakable and Unspeakable in Quantum Mechanics* (Cambridge University Press, Cambridge).

Bell, J. (1987b). Beables for Quantum Field Theory, pp. 173–180 in *Speakable and Unspeakable in Quantum Mechanics* (Cambridge University Press, Cambridge).

Bell, J. (1987c). Introduction to the Hidden-Variable Question, pp. 29–39 in *Speakable and Unspeakable in Quantum Mechanics* (Cambridge University Press, Cambridge).

Bell, J. (1987d). Quantum Mechanics for Cosmologists, pp. 117–138 in *Speakable and Unspeakable in Quantum Mechanics* (Cambridge University Press, Cambridge).

Bell, J. (1987e). *Speakable and Unspeakable in Quantum Mechanics* (Cambridge University Press, Cambridge).

Bell, J. (1990). Against Measurement, pp. 17–31 in *Sixty-Two Years of Uncertainty*, ed. A. Miller (Plenum, New York).

Bell, J. and Clifton, R. (1995). QuasiBoolean Algebras and Simultaneously Definite Properties in Quantum Mechanics, *International Journal for Theoretical Physics* **34**, 2409–2421.

Beller, M. (1990). Born's Probabilistic Interpretation: A Case Study of 'Concepts in Flux', *Studies in History and Philosophy of Science* **21**, 563–588.

Beltrametti, E., and Cassinelli, G. (1981). *The Logic of Quantum Mechanics* (Addison-Wesley, Reading, MA).

Berndl, K., Dürr, D., Goldstein, S., and Zanghì, N. (1996). Nonlocality, Lorentz Invariance, and Bohmian Quantum Theory, *Physical Review A* **53**, 2062–2073.

Berndl, K., and Goldstein, S. (1994). Comment on 'Quantum Mechanics, Local

Realistic Theories, and Lorentz-Invariant Realistic Theories", *Physical Review Letters* **72**, 780.

Bohm, D. (1952). A Suggested Interpretation of the Quantum Theory in Terms of 'Hidden' Variables, I and II, *Physical Review* **85**, 166–193.

Bohm, D. (1953). Comments on an Article of Takabayasi Concerning the Formulation of Quantum Mechanics with Classical Pictures, *Progress of Theoretical Physics* **9**, 273–287.

Bohm, D., and Hiley, B. (1989). Nonlocality and Locality in the Stochastic Interpretation of Quantum Mechanics, *Physics Reports* **172**, 94–122.

Bohm, D., and Hiley, B. (1993). *The Undivided Universe: An Ontological Interpretation of Quantum Theory* (Routledge, London).

Bohm, D., Hiley, B., and Kaloyerou, P. (1987). An Ontological Basis for the Quantum Theory, *Physics Reports* **144**, 321–375.

Bohm, D., and Vigier, J-P. (1954). Model of the Causal Interpretation in Terms of a Fluid with Irregular Fluctuations, *Physical Review Letters* **96**, 208–216.

Born, M. (1926a). Zur Quantenmechanik der Stossvorgänge, *Zeitschrift für Physik* **37**, 863–867. Reprinted and translated in Wheeler and Zurek (1983).

Born, M. (1926b). Quantenmechanik der Stossvorgänge, *Zeitschrift für Physik* **38**, 803–827. Partially reprinted and translated in Ludwig (1968).

Born, M. (1927). Physical Aspects of Quantum Mechanics, *Nature* **119**, 354–357.

Born, M. (1971). *The Born–Einstein Letters* (Walker, New York).

Brown, H. and Anandan, J. (1995). On the Reality of Spacetime Geometry and the Wavefunction, *Foundations of Physics* **25**, 349–360.

Bub, J. (1974). *The Interpretation of Quantum Mechanics* (D. Reidel, Dordrecht).

Bub, J. (1977). Von Neumann's Projection Postulate as a Possibility Conditionalization Rule in Quantum Mechanics, *Journal of Philosophical Logic* **6**, 381–390.

Bub, J. (1995). On the Structure of Quantal Proposition Systems, *Foundations of Physics* **24**, 1261–1279.

Bub, J., and Clifton, R. (1996). A Uniqueness Theorem for Interpretations of Quantum Mechanics, *Studies in History and Philosophy of Modern Physics* **27B**, 181–219.

Bub, J., Clifton, R., and Monton, B. (1997). The Bare Theory Has No Clothes, forthcoming in *Minnesota Studies in Philosophy of Science*, eds. R. Healey and G. Hellman.

Busch, P. and Lahti, P. (1984). On Various Joint Measurements of Position and Momentum Observables, *Physical Review D* **29**, 1634–1646.

Butterfield, J. (1992a). Bell's Theorem: What it Takes, *British Journal for the Philosophy of Science* **43**, 41–83.

Butterfield, J. (1992b). David Lewis Meets John Bell, *Philosophy of Science* **59**, 26–43.

Butterfield, J. (1995). Vacuum Correlations and Outcome Dependence in Algebraic Quantum Field Theory, pp. 768–785 in *Annals of the New York Acadamy of Sciences*, vol. 755, eds. D. Greenberger and A. Zeilinger (The New York Academy of Sciences, New York).

Butterfield, J. (1996a). World's, Minds, and Quanta, *Aristotelian Society Supplemental Volume* **69**, 113–158.

Butterfield, J. (1996b). Whiter the Minds?, *British Journal for the Philosophy of Science* **47**, 200–221.

Butterfield, J., *et al.* (1993a). Parameter Dependence and Outcome Dependence in

Dynamical Models for Statevector Reduction, *Foundations of Physics* **23**, 341–364.

Butterfield, J., *et al.* (1993b). Parameter Dependence in Dynamical Models for Statevector Reduction, *International Journal of Theoretical Physics* **32**, 2287–2304.

Clauser, J., and Horne, M. (1974). Experimental Consequences of Objective Local Theories, *Physical Review D* **10**, 526–535.

Clauser, J., and Shimony, A. (1978). Bell's Theorem: Experimental Tests and Implications, *Reports on Progress in Physics* **41**, 1881–1927.

Clifton, R. (1993). Getting Contextual and Nonlocal Elements-of-Reality the Easy Way, *American Journal of Physics* **61**, 443–447.

Clifton, R. (1995a). Independently Motivating the Kochen–Dieks Modal Interpretation of Quantum Mechanics, *British Journal for the Philosophy of Science* **46**, 33–57.

Clifton, R. (1995b). Making Sense of the Kochen–Dieks 'No-Collapse' Interpretation of Quantum Mechanics Independent of the Measurement Problem, *Annals of the New York Academy of Science* **755**, 570–578.

Clifton, R. and Niemann, P. (1992). Locality, Lorentz Invariance, and Linear Algebra: Hardy's Theorem for Two Entangled Spin-s Particles, *Physics Letters A* **166**, 177–194.

Clifton, R., Redhead, M., and Butterfield, J. (1991). Generalization of the Greenberger-Horne-Zeilinger Algebraic Proof of Nonlocality, *Foundations of Physics* **21**, 149–184.

Cohen-Tannoudji, C., Bernard, D., and Laloë, F. (1977). *Quantum Mechanics* (John Wiley & Sons, New York).

Coleman, S. (1994). Quantum Mechanics in Your Face. Talk delivered at Cambridge University.

Cufaro-Petroni, N. (1992). On the Structure of the Quantum–Mechanical Probability Models, *Foundations of Physics* **22**, 1379–1397.

Cufaro-Petroni, N., *et al.* (1984). Causal Stochastic Interpretation of Fermi-Dirac Statistics in Terms of Distinguishable Non-Locally Correlated Particles, *Physics Letters A* **101**, 4–6.

Curiel, E. (1996). On the Delicacy of Causal Ascription, talk delivered at the APA Eastern Division Meeting (Atlanta).

Cushing, J. (1994a). Locality/Separability: Is This Necessarily a Useful Distinction?, pp. 107–116 in *PSA 1994* vol. I, eds. D. Hull, M. Forbes, and R. Burian (Philosophy of Science Association, East Lansing, MI).

Cushing, J. (1994b). *Quantum Mechanics: Historical Contingency and the Copenhagen Hegemony* (The University of Chicago Press, Chicago).

Dewdney, C., *et al.* (1992). Wave-Particle Dualism and the Interpretation of Quantum Mechanics, *Foundations of Physics* **22**, 1217–1265.

Dewdney, C., Holland, P., and Kyprianidis, A. (1986). What Happens in a Spin Measurement?, *Physics Letters A* **119**, 259–267.

Dewdney, C., Holland, P., and Kyprianidis, A. (1987). A Causal Account of Non-Local Einstein–Podolsky–Rosen Spin Correlations, *Journal of Physics A* **20**, 4717–4732.

Dewdney, C., Holland, P., and Squires, E. (1993). How Late Measurements of Quantum Trajectories Can Fool a Detector, *Physics Letters A* **184**, 6–11.

Dickson, M. (1994a). Wavefunction Tails in the Modal Interpretation, pp. 366–376 in *PSA 1994* vol. I, eds. D. Hull, M. Forbes, and R. Burian (Philosophy of

Science Association, East Lansing, MI).

Dickson, M. (1994b). What is Preferred about the Preferred Basis?, *Foundations of Physics* **25**, 423–441.

Dickson, M. (1995a). Is There *Really* No Projection Postulate in the Modal Interpretation?, *British Journal for the Philosophy of Science* **46**, 167–188.

Dickson, M. (1995b). Faux-Boolean Algebras, Classical Probability, and Determinism, *Foundations of Physics Letters* **8**, 231–242.

Dickson, M. (1995c). Faux-Boolean Algebras and Classical Models, *Foundations of Physics Letters* **8**, 401–415.

Dickson, M. (1995d). Probability and Nonlocality: Determinism Versus Indeterminism in Quantum Mechanics, Ph.D. Dissertation (University of Notre Dame).

Dickson, M. (1996a). Determinism and Locality in Quantum Systems, *Synthese* **107**, 52–82.

Dickson, M. (1996b). Logical Foundations for Modal Interpretations *Philosophy of Science* **63 (Supp)**, 322–329.

Dickson, M. (1996c). Antidote or Theory? *The Undivided Universe* and *The Quantum Theory of Motion*, *Studies in History and Philosophy of Modern Physics* **27B** 229–238.

Dickson, M. (1996d). Is The Bohm Theory Local? pp. 321–330 in *Bohmian Mechanics and Quantum Theory: An Appraisal*, eds. J. Cushing, A. Fine, and S. Goldstein (Kluwer, Dordrecht)

Dickson, M. (1997). On the Plurality of Dynamics, forthcoming in *Minnesota Studies in Philosophy of Science*, eds. R. Healey and G. Hellman.

Dickson, M. and Clifton, R. (1997). Lorentz-Invariance in Modal Interpretations, forthcoming in *The Modal Interpretation of Quantum Mechanics*, eds. D. Dieks and P. Vermaas (Kluwer, Dordrecht).

Dieks, D. (1988a). The Formalism of Quantum Theory: An Objective Description of Reality?, *Annalen der Physik* **7**, 174–190.

Dieks, D. (1988b). Special Relativity and the Flow of Time *Philosophy of Science* **55**, 456–460.

Dieks, D. (1989). Resolution of the Measurement Problem Through Decoherence of the Quantum State, *Physics Letters A* **142**, 439–446.

Dieks, D. (1994). The Modal Interpretation of Quantum Mechanics, Measurements and Macroscopic Behavior, *Physical Review A* **49**, 2290–2300.

Dieks, D. (1997). Preferred Factorizations and Consistent Property Attribution, forthcoming in *Minnesota Studies in Philosophy of Science*, eds. R. Healey and G. Hellman.

Diosi, L. (1992). Quantum Measurement and Gravity for Each Other, pp. 299–304 in *Quantum Chaos—Quantum Measurement*, eds. P. Cvitanovic, I. Percival, and A. Wirzba (Kluwer, Dordrecht).

Dummett, M. (1976). Is Logic Empirical?, pp. 45–68 in *Contemporary British Philosophy*, ed. H. Lewis (George Allen and Unwin, London).

Dürr, D., *et al.* (1993). Comment on: Surrealistic Bohm Trajectories, *Zeitschrift für Naturforsch* **48a**, 1261–1262.

Dürr, D., Goldstein, S., and Zanghi, N. (1992). Quantum Chaos, Classical Randomness. and Bohmian Mechanics, *Jounal of Statistical Physics* **68**, 259–270.

Dürr, D., Goldstein, S., and Zanghi, N. (1996). Bohmian Mechanics as the Foundation of Quantum Mechanics, pp. 21–44 in *Bohmian Mechanics and*

Quantum Theory: An Appraisal, eds. J. Cushing, A. Fine, and S. Goldstein (Kluwer, Dordrecht).

Earman, J. (1986). *A Primer on Determinism* (D. Reidel, Dordrecht).

Elby, A. (1993). Why 'Modal' Interpretations of Quantum Mechanics Don't Solve the Measurement Problem, *Foundations of Physics Letters* **6**, 5–19.

Elby, A. (1994). The 'Decoherence' Approach to the Measurement Problem in Quantum Mechanics, in *PSA 1994* vol. I, eds. D. Hull, M. Forbes, and R. Burian (Philosophy of Science Association, East Lansing, MI).

Englert, B-G., *et al.* (1992). Surrealistic Bohm Trajectories, *Zeitschrift für Naturforschung* **47a**, 1175–1186.

Everett, H. (1957). "Relative State" Formulation of Quantum Mechanics, *Reviews of Modern Physics* **29**, 454–462.

Fevrier, P. (1937). Les Relations d'Incertitude de Heisenberg et la Logic', *Comptes Rendus des Séances de l'Académie des Sciences* **204**, 481–483.

Fine, A. (1970). Insolubility of the Measurment Problem, *Physical Review D* **24**, 1–37.

Fine, A. (1982a). Hidden Variables, Joint Probability, and the Bell Inequalities, *Physical Review Letters* **48**, 291–5.

Fine, A. (1982b). Joint Distributions, Quantum Correlations, and Commuting Observables, *Journal of Mathematical Physics* **23**, 1306–10.

Fleming, G. (1989). Lorentz Invariant State Reduction, and Localization, pp. 112–126 in *PSA 1988* vol. II, eds. A. Fine and J. Leplin (Philosophy of Science Association, East Lansing, MI).

Fleming, G., and Bennett, H (1989). Hyperplane Dependence in Relativistic Quantum Mechanics, *Foundations of Physics* **19**, 231–267.

Friedman, M. and Putnam, H. (1978). Quantum Logic, Conditional Probability and Inference, *Dialectica* **32**, 305–315.

Gell-Mann, M. (1994). *The Quark and the Jaguar* (W. H. Freedman and Co., New York).

Gell-Mann, M., and Hartle, J. (1990). Quantum Mechanics in the Light of Quantum Cosmology, pp. 425–458 in *Complexity, Entropy, and the Physics of Information*, ed. W. Zurek (Addison-Wesley, New York).

Gell-Mann, M. and Hartle, J. (1993). Classical Equations for Quantum Systems, *Physics Review A* **47**, 3345–3382.

Ghirardi, G-C., and Grassi, R (1994). Outcome Predictions and Property Attribution: the EPR Argument Reconsidered, *Studies in History and Philosophy of Modern Physics* **25**, 397–423.

Ghirardi, G-C., Grassi, R., and Pearle, P. (1990a). Continuous-Spontaneous-Reduction Model Involving Gravity, *Physical Review A* **42**, 1057–1065.

Ghirardi, G-C., Grassi, R., and Pearle, P. (1990b). Relativistic Dynamical Reduction Models: General Framework and Examples, *Foundations of Physics* **20**, 1271–1316.

Ghirardi, G-C., and Pearle, P. (1990). Elements of Physical Reality, Nonlocality and Stochasticity in Relativistic Dynamical Reduction Models, pp. 35–47 in *PSA 1990* vol. II, eds. A. Fine, M. Forbes, and L. Wessels, L. (Philosophy of Science Association, East Lansing, MI).

Ghirardi, G-C., Pearle, P., and Rimini, A. (1990). Markov Processes in Hilbert Space and Continuous Spontaneous Localization of Systems of Identical Particles, *Physical Review A* **42**, 78–89.

Ghirardi, G-C., Rimini, A., and Weber, T. (1986). Unified Dynamics for Microscopic and Macroscopic Systems, *Physical Review D* **34**, 470–491.

Gisin, N. (1984). Quantum Measurement and Stochastic Processes, *Physical Review Letters* **52**, 1657–1660.

Gisin, N. (1989). Stochastic Quantum Dynamics and Relativity, *Helvetica Physica Acta* **62**, 363–371.

Gleason, A. (1957). Measures on the Closed Subspaces of a Hilbert Space, *Journal of Mathematics and Mechanics* **6**, 885–893.

Goldstein, S. (1996a). Naïve Realism About Operators, talk delivered to the Philosophy of Science Association Bienniel Meeting (New Orleans).

Goldstein, S. (1996b). Bohmian Mechanics and the Quantum Revolution, *Synthese* **107**, 145–165.

Greenberger, D. M., Horne, M. A., Shimony, A., and Zeilinger, A. (1990). Bell's Theorem Without Inequalities, *American Journal of Physics* **58**, 1131–43.

Greenberger, D. M., Horne, M. A., and Zeilinger, A. (1989). Going Beyond Bell's Theorem, pp. 69–72 in *Bell's Theorem, Quantum Theory, and Conceptions of the Universe*, ed. M. Kafatos (Kluwer, Dordrecht).

Griffiths, R. (1984). Consistent Histories and the Interpretation of Quantum Mechanics, *Journal of Statistical Physics* **36**, 219–272.

Griffiths, R. (1993). The Consistency of Consistent Histories, *Foundations of Physics* **23**, 1601–1610.

Groenewold, H. (1957). Objective and Subjective Aspects of Statistics in Quantum Description, pp. 197–208 in *Observation and Interpretation in the Philosophy of Physics*, ed. S. Körner (Butterworths Scientific Publications, London).

Gudder, S. (1979). *Stochastic Methods in Quantum Mechanics* (North Holland, New York).

Gudder, S. (1988). *Quantum Probability* (Academic Press, Boston).

Halmos, P. (1969). Two Subspaces, *Transactions of the American Mathematical Society* **144**, 381–389.

Hardy, L. (1992). Quantum Mechanics, Local Realistic Theories, and Lorentz-Invariant Realistic Theories, *Physical Review Letters* **68**, 2981–2984.

Hardy, L. (1994). Reply to Comment on 'Quantum Mechanics, Local Realistic Theories, and Lorentz-Invariant Realistic Theories", *Physical Review Letters* **72**, 781.

Hardy, L. and Squires, E. (1992). On the Violation of Lorentz-Invariance in Deterministic Hidden-Variables Interpretations of Quantum Theory, *Physics Letters A* **168**, 169–173.

Healey, R. (1989). *The Philosophy of Quantum Mechanics: An Interactive Interpretation* (Cambridge University Press, Cambridge).

Healey, R. (1991). Holism and Nonseparability, *Journal of Philosophy* **88**, 393–421.

Healey, R. (1994). Nonseparable Processes and Causal Explanation, *Studies in History and Philosophy of Modern Physics* **25**, 337–374.

Healey, R. (1997). "Modal" Interpretations, Decoherence and the Quantum Measurement Problem, forthcoming in *Minnesota Studies in Philosophy of Science*, eds. R. Healey and G. Hellman.

Heilbron, J. (1988). The Earliest Missionaries of the Copenhagen Spirit, pp. 201–233 in *Science in Reflection*, ed. E. Ullmann-Margalit (Kluwer, Dordrecht).

Hellwig, K., and Crause, K. (1970). Formal Description of Measurements in Local Quantum Theory, *Physical Review D* **1**, 566–571.

Holland, P. (1992). The Dirac Equation in the de Broglie–Bohm Theory of Motion, *Foundations of Physics* **22**, 1287–1301.

Holland, P. (1993). *The Quantum Theory of Motion: An Account of the de Broglie–Bohm Causal Interpretation* (Cambridge University Press, Cambridge).

Howard, D. (1985). Einstein on Locality and Separability, *Studies in the History and Philosophy of Science* **16**, 171–201.

Howard, D. (1989). Holism, Separability, and the Metaphysical Implications of the Bell Experiments, pp. 224–253 in *Philosophical Consequences of Quantum Theory*, eds. J. Cushing and E. McMullin (University of Notre Dame Press, Notre Dame, IN).

Hughes, R. (1989). *The Structure and Interpretation of Quantum Mechanics* (Harvard University Press, Cambridge, MA).

Jammer, M. (1966). *The Conceptual Development of Quantum Mechanics* (McGraw-Hill, New York).

Jarrett, J. (1984). On the Physical Significance of the Locality Conditions in the Bell Arguments, *Noûs* **18**, 569–589.

Jarrett, J. (1986). Does Bell's Theorem Apply to Theories That Admit Time-Dependent States?, *Annals of the New York Academy of Sciences* **480**, 428–437.

Jarrett, J. (1989). Bell's Theorem: A Guide to the Implications, pp. 60–79 in *Philosophical Consequences of Quantum Theory*, eds. J. Cushing and E. McMullin (University of Notre Dame Press, Notre Dame, IN).

Jones, M., and Clifton, R. (1993). Against Experimental Metaphysics, pp. 295–316 in *Midwest Studies in Philosophy* vol. 18, eds. P. French, T. Uehling, and H. Wettstein (University of Notre Dame Press, Notre Dame, IN).

Jordan, P. (1927). Philosophical Foundations of Quantum Theory, *Nature* **119**, 566–569.

Kent, A. (1995). A Note on Schmidt States and Consistency, *Physics Letters* **196**, 313.

Kochen, S. (1985). A New Interpretation of Quantum Mechanics, pp. 151–170 in *Symposium on the Foundations of Modern Physics*, eds. P. Lahti and P. Mittelstaedt (World Scientific, Singapore).

Kochen, S., and Specker, E. (1967). The Problem of Hidden Variables in Quantum Mechanics, *Journal of Mathematics and Mechanics* **17**, 59–87.

Krips, H. (1987). *The Metaphysics of Quantum Theory* (Clarendon Press, Oxford).

Laudisa, F. (1995). Einstein, Bell, and Nonseparable Realism, *British Journal for the Philosophy of Science* **46**, 309–329.

Lewis, D. (1973). *Counterfactuals* (Blackwell, Oxford).

Lewis, D. (1986). Causation, pp. 159–172 in *Philosophical Papers* vol. II. (Oxford University Press, Oxford).

Lockwood, M. (1996). 'Many Minds' Interpretations of Quantum Mechanics, *British Journal for the Philosophy of Science* **47**, 159–188.

Ludwig, G. (1968). *Wave Mechanics* (Oxford University Press, Oxford).

Mackman, S., and Squires, E. (1995). Lorentz Invariance and the Retarded Bohm Model, *Foundations of Physics* **25**, 391–397.

Malament, D. (1996). In Defense of Dogma, pp. 1–10 in *Perspectives on Quantum Reality*, ed. R. Clifton (Kluwer, Dordrecht).

Margenau, H. (1963). Measurement and Quantum States, *Philosophy of Science* **11**, 1–16.

Marshall, T., Santos, E., and Selleri, F. (1983). Local Realism Has Not Been Refuted by Atomic Cascade Experiments, *Physics Letters A* **98**, 5–9.

Maudlin, T. (1994). *Quantum Non-Locality and Relativity* (Blackwell, Oxford).

Maxwell, N. (1985). Are Probabilism and Special Relativity Incompatible?, *Philosophy of Science* **52**, 23–43.

Maxwell, N. (1988). Discussion: Are Probabilism and Special Relativity Compatible?, *Philosophy of Science* **55**, 640–645.

Mermin, D. (1990). Simple Unified Form for the Major No-Hidden-Variables Theorems, *Physical Review Letters* **65**, 3373–3376.

Mermin, D. (1994). Hidden Variables and the Two Theorems of John Bell, *Reviews of Modern Physics* **65**, 803–815.

Nelson, E. (1967). *Dynamical Theories of Brownian Motion* (Princeton University Press, Princeton, NJ).

Omnés, R. (1989). Logical Reformulation of Quantum Mechanics, *Journal of Statistical Physics* **53**, 893–932.

Omnés, R. (1994). *The Interpretation of Quantum Mechanics* (Princeton University Press, Princeton, NJ).

Pagonis, C., Redhead, M. L. G., and Clifton, R. K. (1991). Breakdown of Quantum Nonlocality in the Classical Limit, *Physics Letters A* **155**, 441–444.

Pauli, W. (1927). Über Gasentartung und Paramagnetismus, *Zeitschrift für Physik* **43**, 81–102.

Pearle, P. (1989). Combining Stochastic Dynamical State-Vector Reduction with Spontaneous Localization, *Physical Review A* **39**, 2277–2289.

Pearle, P. (1990). Toward a Relativistic Theory of Statevector Reduction, pp. 193–214 in *Sixty-Two Years of Uncertainty*, ed. A. Miller (Plenum, New York).

Pearle, P. (1992a). Relativistic Model for Statevector Reduction, pp. 283–297 in *Quantum Chaos—Quantum Measurement*, ed. P. Cvitanovic̀ (Kluwer, Dordrecht).

Pearle, P. (1992b). Reality Checkpoint, pp. 211–227 in *Acta Encyclopedia* (Instituto della Enciclopedia Italiana, Rome).

Penrose, R. (1996). On Gravity's Role in Quantum State Reduction, *General Relativity and Gravitation* **28**, 581–609.

Peres, A. (1978). Unperformed Experiments Have No Results, *American Journal of Physics* **46**, 745–747.

Peres, A. (1994). *Quantum Theory: Concepts and Methods* (Kluwer, Dordrecht).

Primas, H. (1990). Induced Nonlinear Time Evolution of Open Quantum Objects, pp. 259–280 in *Sixty-Two Years of Uncertainty*, ed. A. Miller (Plenum, New York).

Prugovečki, E. (1984). *Stochastic Quantum Mechanics and Quantum Spacetime* (D. Reidel, Dordrecht).

Putnam, H. (1957). Three-Valued Logic, *Philosophical Studies* **8**, 73–80.

Putnam, H. (1967). Time and Physical Geometry, *The Journal of Philosophy* **64**, 240–247.

Putnam, H. (1969). Is Logic Empirical?, *Boston Studies in the Philosophy of Science* **5**, 199–215.

Putnam, H. (1974). How To Think Quantum Logically, *Synthese* **29**, 55–61.

Redhead, M. (1987). *Incompleteness, Nonlocality, and Realism* (Clarendon Press, Oxford).

Reed, M., and Simon, B. (1979). *Methods of Modern Mathematical Physics* (Academic Press, San Diego, CA).

Reichenbach, H. (1944). *Philosophic Foundations of Quantum Mechanics* (University of California Press, Los Angeles).

Reichenbach, H. (1956). *The Direction of Time* (University of California Press, Berkeley, CA).

Rietdjik, C. (1966). A Rigorous Proof of Determinism Derived from the Special Theory of Relativity, *Philosophy of Science* **33**, 341–344.

Rietdjik, C. (1976). Special Relativity and Determinism, *Philosophy of Science* **43**, 598–609.

Sakurai, J. (1967). *Advanced Quantum Mechanics* (Addison-Wesley, Reading, MA).

Salmon, W. (1984). *Scientific Explanation and the Causal Structure of the World* (Princeton University Press, Princeton, NJ).

Saunders, S. (1993). Decoherence, Relative States, and Evolutionary Adaptation, *Foundations of Physics* **23**, 1553–1595.

Schrödinger, E. (1926). Quantisierung als Eigenwertproblem, *Annalen der Physik* **79**, 361–376.

Schrödinger, E. (1935a). Die gegenwartige Situation in der Quantenmechanik, *Naturwissenschaften* **23**, 807–812; 823–828; 844–849.

Schrödinger, E. (1935b). Discussion of Probability Relations Between Separated Systems, *Proceedings of the Cambridge Philosophical Society* **31**, 555–563.

Schrödinger, E. (1952). Are There Quantum Jumps?, *British Journal for the Philosophy of Science* **3**, 109–123; 233–247.

Shimony, A. (1986). Events and Processes in the Quantum World, pp. 182–203 in *Quantum Concepts in Space and Time*, eds. R. Penrose and C. Isham (Clarendon Press, Oxford).

Stapp, H. (1992). Noise-Induced Reduction of Wave Packets and Faster-Than-Light Influences, *Physical Review A* **46**, 6860–6866.

Stein, H. (1970). On Einstein–Minkowski Space-Time, *The Journal of Philosophy* **65**, 5–23.

Stein, H. (1991). On Relativity and Openness of the Future, *Philosophy of Science* **58**, 147–167.

Suppes, P., and Zanotti, M. (1976). On the Determinism of Hidden Variable Theories with Strict Correlation and Conditional Statistical Independence of Observables, pp. 445–455 in *Logic and Probability in Quantum Mechanics*, ed. P. Suppes (D. Reidel, Dordrecht).

Svetlichny, G., Redhead, M., Brown, H., and Butterfield, J. (1988). Do the Bell Inequalities Require the Existence of Joint Probability Distributions?, *Philosophy of Science* **55**, 387–401.

Teller, P. (1984). The Projection Postulate: A New Perspective, *Philosophy of Science* **51**, 369–395.

Teller, P. (1989). Relativity, Relational Holism, and the Bell Inequalities, pp. 208–223 in *Philosophical Consequences of Quantum Theory*, eds. J. Cushing and E. McMullin (University of Notre Dame Press, Notre Dame, IN).

Valentini, A. (1991a). Signal-Locality, Uncertainty, and the Subquantum *H*-Theorem. I, *Physics Letters A* **156**, 5–11.

Valentini, A. (1991b). Signal-Locality, Uncertainty, and the Subquantum *H*-Theorem. II, *Physics Letters A* **158**, 1–8.

van Fraassen, B. (1979). Hidden Variables and the Modal Interpretation of Quantum Theory, *Synthese* **42**, 155–165.

van Fraassen, B. (1980). *The Scientific Image* (Oxford University Press, Oxford).

van Fraassen, B. (1981). A Modal Interpretation of Quantum Mechanics, pp. 229–258 in *Current Issues in Quantum Logic*, eds. E. Beltrametti and B. van Fraassen (Plenum, New York).

van Fraassen, B. (1982). Rational Belief and the Common Cause Principle, pp. 193–209 in *What? Where? When? Why?*, ed. R. McLaughlin (D. Reidel, Dordrecht).

van Fraassen, B. (1990). The Modal Interpretation of Quantum Mechanics, pp. 440–460 in *Symposium on the Foundations of Modern Physics 1990*, eds. P. Lahti and P. Mittelstaedt (World Scientific, Singapore).

van Fraassen, B. (1991). *Quantum Mechanics: An Empiricist View* (Clarendon Press, Oxford).

Vermaas, P. (1996a). Unique Transition Probabilities and the Modal Interpretation, *Studies in the History and Philosophy of Modern Physics*, **27B**, 133–159.

Vermaas, P. (1996b). A No-Go Theorem for Joint Property Ascriptions in the Modal Interpretation of Quantum Mechanics, unpublished manuscript.

Vermaas, P., and Dieks, D. (1995). The Modal Interpretation of Quantum Mechanics and Its Generalization to Density Operators, *Foundations of Physics* **25**, 145–158.

Vink, J. (1993). Quantum Mechanics in Terms of Discrete Beables, *Physical Review A* **48**, 1808–1818.

von Neumann, J. (1955). *Mathematical Foundations of Quantum Mechanics* (Princeton University Press, Princeton, NJ).

Wessels, L. (1981). What Was Born's Statistical Interpretation?, pp. 187–200 in *PSA 1980* vol. II, eds. P. Asquith and R. Giere (Philosophy of Science Association, East Lansing, MI).

Wheeler, J., and Zurek, W., eds. (1983). *Quantum Theory and Measurement* (Princeton University Press, Princeton).

Wigner, E. (1983). Interpretation of QUantum Mechanics, pp. 260–314 in *Quantum Theory and Measurement*, eds. J. Wheeler and W. Zurek (Princeton University Press, Princeton, NJ).

Zimba, J. and Clifton, R. (1997). Valuations on Functionally Closed Sets of Quantum-Mechanical Observables and Von Neumanns No-Hidden-Variables Theorem, forthcoming in *The Modal Interpretation of Quantum Mechanics*, eds. D. Dieks and P. Vermaas (Kluwer, Dordrecht).

Zurek, W. (1993). Preferred States, Predictability, Classicality, and the Environment-Induced Decoherence, *Progress in Theoretical Physics* **89**, 281–312.

Index

242

ated in the United States
Bookmasters